Valuing Development, Environment and Conservation

Policy-makers are increasingly trying to assign economic values to areas such as ecologies, the atmosphere, even human lives. These new values, assigned to areas previously considered outside of economic systems, often act to qualify, alter or replace former non-pecuniary values. *Valuing Development, Environment and Conservation* looks to explore the complex interdependencies, contradictions and trade-offs that can take place between economic values and the social, environmental, political and ethical systems that inform non-monetary valuation processes.

Using rich empirical material, the book explores the processes of valuation, their components, calculative technologies, and outcomes in different social, ecological and conservation domains. The book gives reasons for why economic calculation tends to dominate in practice, but also presents new insights on how the disobedient materiality of things and the ingenuity of human and non-human agencies can combine and frustrate the dominant economic models within calculative processes.

This book highlights the tension between, on the one hand, a dominant model that emphasises technical and 'universalising' criteria, and on the other hand, valuation practice in specific local contexts which is more likely to negotiate criteria that are plural, incommensurable and political. This book is perfect for researchers and students within development studies, environment, geography, politics, sociology and anthropology who are looking for new insights into how processes of valuation take place in the 21st century, and with what consequential outcomes.

Sarah Bracking is Professor of Climate and Society in the School of Global Affairs, King's College London, UK.

Aurora Fredriksen is a Lecturer in Human Geography at the University of Manchester, UK.

Sian Sullivan is Professor of Environment and Culture, Bath Spa University, UK.

Philip Woodhouse is Professor of Environment and Development, University of Manchester, UK.

Routledge Explorations in Development Studies

This Development Studies series features innovative and original research at the regional and global scale. It promotes interdisciplinary scholarly works drawing on a wide spectrum of subject areas, in particular politics, health, economics, rural and urban studies, sociology, environment, anthropology, and conflict studies.

Topics of particular interest are globalization; emerging powers; children and youth; cities; education; media and communication; technology development; and climate change.

In terms of theory and method, rather than basing itself on any orthodoxy, the series draws broadly on the tool kit of the social sciences in general, emphasizing comparison, the analysis of the structure and processes, and the application of qualitative and quantitative methods.

Learning and Forgetting in Development NGOs
Insights from Organisational Theory
Tiina Kontinen

Information Communication Technology and Poverty Alleviation
Promoting Good Governance in the Developing World
Jack J. Barry

Engendering Transformative Thinking and Practice in International Development
Gillian Fletcher

Valuing Development, Environment and Conservation
Creating Values that Matter
Edited by Sarah Bracking, Aurora Fredriksen, Sian Sullivan and Philip Woodhouse

Valuing Development, Environment and Conservation

Creating Values that Matter

Edited by
Sarah Bracking, Aurora Fredriksen,
Sian Sullivan and Philip Woodhouse

Routledge
Taylor & Francis Group
LONDON AND NEW YORK

First published 2019
by Routledge
2 Park Square, Milton Park, Abingdon, Oxon OX14 4RN

and by Routledge
52 Vanderbilt Avenue, New York, NY 10017, USA

First issued in paperback 2020

Routledge is an imprint of the Taylor & Francis Group, an informa business

British Library Cataloguing-in-Publication Data
A catalogue record for this book is available from the British Library

Library of Congress Cataloging-in-Publication Data
Names: Bracking, Sarah, 1968– editor.
Title: Valuing development, environment and conservation : creating values that matter / edited by Sarah Bracking, Aurora Fredriksen, Sian Sullivan and Philip Woodhouse.
Description: Abingdon, Oxon ; New York, NY : Routledge, 2019. | Series: Routledge explorations in development studies | Includes bibliographical references and index.
Identifiers: LCCN 2018024446| ISBN 9781138080515 (hardback) | ISBN 9781315113463 (ebook)
Subjects: LCSH: Environmental protection—Economic aspects. | Conservation of natural resources—Economic aspects. | Economic development. | Value—Social aspects.
Classification: LCC HC79.E5 V26128 2019 | DDC 333.7—dc23
LC record available at https://lccn.loc.gov/2018024446

ISBN 13: 978-0-367-66500-5 (pbk)
ISBN 13: 978-1-138-08051-5 (hbk)

Typeset in Times New Roman
by codeMantra

Contents

Figures and tables

Figures

Tables

Contributors

Sarah Bracking is Professor of Climate and Society in the School of Global Affairs, King's College London, UK, and formerly Senior Professor and Department of Science and Technology and National Research Foundation-funded South African Research Chair (SARCHi) at the University of KwaZulu-Natal, South Africa, and Research Director of the Leverhulme Centre for the Study of Value, University of Manchester, UK.

Jonas Amtoft Bruun is a researcher on the Climate Justice Research Project at Dartmouth College, USA, and holds a PhD from the Global Development Institute at the University of Manchester. His research interests span international politics, sustainability, and stakeholder engagement, and his recent work explores civil society participation in the governing of global climate finance.

Louise Emily Carver is a human geographer exploring the science–policy interface of biodiversity conservation. She conducts institutional ethnographies, documenting practices of valuation in the green economy and increasingly their interface with digital technologies.

Aurora Fredriksen is a Lecturer in Human Geography at the University of Manchester, UK, and formerly the Assistant Director of the Leverhulme Centre for the Study of Value. Her research explores the ways that science and technology mediate practices of care across difference, with a focus on more-than-human agencies and worldings.

Fortunate Machingura is a Research Fellow in the Growth, Poverty and Inequality Programme at the Overseas Development Institute in London. She is also an Honorary Research Fellow at the University of Manchester's Global Development Institute. Her research spans health policy and health systems studies, with particular interests in health planning and politics, HIV/AIDS, Health Management Information Systems, community participation, Sustainable Development Goals and social determinants of health and how these intersect with health outcomes.

Mike Muller is a Visiting Adjunct Professor at the Witwatersrand University School of Governance, South Africa. A professional engineer, he was a Commissioner of South Africa's National Planning Commission (2010–15) and Director-General of South Africa's Department of Water Affairs and Forestry (1997–2005). As chair of the World Economic Forum's Agenda Council on Water (2012–14), he promoted a review of values-based approaches to the achievement of water security.

Rebecca Peters is currently a DPhil candidate at the University of Oxford, UK, in the School of Geography and the Environment, where she is an EPA Marshall Scholar. Her scholarship and advocacy focus on the intersection of water security, environmental health, and social justice. She has professional experience working on water projects in over 10 countries including Bolivia, Mexico, South Africa, and China. Rebecca holds an MSc in Poverty and Development Economics from the University of Manchester and an MSc in Water Science and Governance from King's College London.

Sian Sullivan is an environmental anthropologist and political ecologist working at the intersections between culture, nature and finance. Currently she is Professor of Environment and Culture at Bath Spa University, UK, and Principal Investigator for two UK Arts and Humanities Research Council projects ('Future Pasts' and 'Disrupted Histories, Recovered Pasts').

Robert Watt is a Lecturer in Politics at the University of Manchester, UK. His primary research interests are in the international politics of climate change, and his recent work examines the cultural political economy of carbon markets.

Philip Woodhouse is Professor of Environment and Development at the University of Manchester, UK. He trained as an agricultural scientist and has worked for the past 30 years on the political economy of agricultural productivity and land and water governance, principally in sub-Saharan Africa. He co-authored *African Enclosures: The social dynamics of wetlands in drylands* (2000) and *Water Resources and Development* (2011).

Acknowledgements

This book is a result of research carried out within the Leverhulme Centre for the Study of Value (LCSV) using funding from the Leverhulme Trust under their research programme grant scheme, award no. RP2012-V-041. Within this research programme titled *Human, non-human and environmental values: an impossible frontier?* we carried out case studies under five separate research themes, each concerned with the ways people and the non-human world are valued in the humanitarian, development, environmental and agricultural fields. All of our cases explored how calculative rationality, accounting 'devices', notional values and value framings are increasingly incorporating evermore entities into socially articulated markets and spaces as well as the effects these have on the wellbeing of people and the non-human environment.

The LCSV was established in 2012 by Professor Sarah Bracking, with the generous support of the Leverhulme Trust, to whom we are extremely grateful. The LCSV supported an international and inter-institutional team of researchers, including project Co-Investigators, Research Associates and postgraduate students based at the University of Manchester and at Birkbeck, University of London, as well as research affiliates at the University of KwaZulu-Natal in South Africa, Bath Spa University in the UK and the University of Virginia in the United States. Our interdisciplinary team of researchers combined a diverse range of theoretical perspectives, methodological approaches and field experiences to bear on the study of value. Sarah Bracking was the Principal Investigator and award holder, while Sian Sullivan and Philip Woodhouse were Co-Investigators of the programme and leaders of their respective research projects and teams, with administrative, managerial and intellectual responsibilities shared in a collective structure. Patrick Bond and James Igoe were internationally affiliated, and Dan Brockington playing an important role as the initial adviser to the project documentation. They were joined by Research Associates Aurora Fredriksen and Elisa Greco, and Doctoral students Jonas Amtoft Bruun, Louise Emily Carver, Fortunate Machingura, Rachael Morgan and Robert Watt. More information about LCSV and our working paper series can be found on our website, thestudyofvalue.org. The authors of this volume thank Pauline Nelson for her copy-editing of the chapters. There are many persons, and things, who contributed to the success of the programme of research and lest we omit any accidently we give only a collective vote of appreciation here.

The Leverhulme Trust was established in 1925 under the Will of the first Viscount Leverhulme. It is one of the largest all-subject providers of research funding in the UK, distributing funds of some £60 million every year. For further information about the schemes that the Leverhulme Trust fund visit their website at www.leverhulme.ac.uk/www.twitter.com/LeverhulmeTrust

1 Introducing values that matter

Sarah Bracking, Aurora Fredriksen, Sian Sullivan and Philip Woodhouse

Introduction

In 2010 the UK government mandated that all government spending demonstrate 'value for money', a valuation practice that took on particular salience in the Department for International Development (DfID) and has since been taken up by other big donors as a means for valuing international development aid. In 2015 a contract for a biodiversity offset package in England was priced at £120,000 for 25.52 biodiversity units, including 6.36 units of biodiversity value contributed by football pitches and their margins. In the same year, in South Africa, the Minister of Tourism was keen to promote rhinoceros conservation in terms of these animals 'paying to stay'. Simultaneously, multiple new coal mines in Somkhele used the iMfolozi watershed for the washing of millions of tonnes of coal, despite perennial water shortages in neighbouring Mpukunyoni communities and the needs of wildlife – including rhinoceroses – in the adjacent Hluhluwe-iMfolozi national heritage park. The ensuing damage to the watershed was to be 'offset' by a Black Economic Empowerment (BEE) holding assigned to the local political elite. In the following year the exchange value of carbon credits – Certified Emission Reduction (CER) units tradable through the Clean Development Mechanism (CDM) with the aim of encouraging reduced carbon emissions globally – had fallen from a high of €16.8/tCO_2e in 2008 to almost nothing. Nonetheless, by mid-2017, the newly created flagship Green Climate Fund (GCF) calculated that 128 million persons globally were 'anticipated' to have 'increased resilience' because it was generating 978 million tonnes of avoided CO_2 emissions from its USD10 billion in planned and on-going investments (GCF 2017).

As the above examples suggest, new markets and commodities are being created in a number of key policy arenas that are monetising, and thereby imputing a particular type of value to, a number of previously unpriced, but not necessarily unvalued, things. Carbon, 'ecosystem services', commonly held land, and human lives are being (re)valued in economic terms. Assigning prices to units of beyond-human nature so as to make these 'count' – either as 'ecosystem services' or in other market or market-like schemes for valuing living entities and ecological processes – has been a particularly visible arena for the rise of economic valuations (e.g. Lohmann 2009; Adler 2013; Helm 2015; ten Kate and Crowe 2014). But this is not the only place where newly economic values are on the rise. New markets for land in the global South evidence the spread of existing market valuation practices to different geographic areas, while the rise of water markets across the globe look with varying degrees of success (Bakker 2003) for ways of framing and making water as a privatised commodity. Even in international development and public health, areas where economic value has long been an essential measure of worth, we are seeing new uses of economic valuation as a tool for demonstrating the worth of what would have previously been left to the (non-economic) judgements of experts, namely: programming, spending decisions and human lives.

Simultaneously, however, an intensification of pricing practices to indicate values that will incentivise particular decisions is mirrored by proliferating critique on ethical grounds. Popularised recently through texts such as philosopher Michael Sandel's (2013) *What Money Can't Buy: The Moral Limits of Markets*, a diversity of views consider that multiple social and ecological choices should not be left to market valuation alone, that intrinsic values cannot be revealed through price mechanisms, and that modern market preferences frequently are unable to recognise prior value practices that confer care (see, for example, Foster 1997; O'Neill 2007; Martin et al. 2013). These perspectives echo Oscar Wilde's (2005[1892]) pithy observation of the late 1800s that a cynic 'knows the price of everything and the value of nothing'. Marketisation processes can also suggest an interchangeability of assets or notional offsets, assuming, for example, that losses of specific emplaced populations or habitats can be easily restored or provided for elsewhere. This assumption obscures the uniqueness of each location and ignores the reality that scope for restitutions of a like kind is shrinking. With only two estuarine river sandbanks left in the Republic of South Africa for spawning frogs and fish, for example, the company behind a proposed port expansion in Durban argued that it could destroy one of these remaining sites because lost populations and habitats could be compensated for with an offset site somewhere else. That the particularly affected species would have to travel around the Cape to get to the proposed offset site was ignored. Nevertheless, it seems that the fetishised power of value as capital – usually a synonym for money – even has many self-avowed radicals convinced that ecological justice can arise as a consequence of the clarifications of 'wealth' afforded by such calculations (e.g. Bond 2013).

Ethically and cross-culturally grounded critiques of the increasing hegemony of economic valuations in decision-making provide a much-needed counter to narratives positing that 'there is no alternative' to the progressive economisation of societal interactions. At the same time, such critiques tend to accept economic value on its own terms: as a type of value that can be rationally assessed and assigned, however detrimentally or unethically. In short, economic valuation is seen to confer meaning to the distribution, protection and disposal of things and people without reflective or normative thinking being in much evidence. This is either around the ethics of the assignment and its consequences or, more foundationally, about the ability of economic valuation to be rational, accurate or relevant in the first place. The situation reflects the power of economics, statistics and numeric judgements more generally to confer an appearance of robustness, fixity and certainty (e.g. Poovey 1998; Power 1997, 2003; Porter 1995). That the public is so accepting of economic forms of valuation demonstrates the domineering power of the field of economics vis-à-vis other knowledge paradigms in modern culture (Kjellberg et al. 2013). Indeed, money and finance hold an almost mythic status as measures of value (Guyer 2004), despite the many complexities they remove from view.

The works collected in this volume instead challenge the dominance of the economic as the measure of all value. They do this by examining how economic valuation technologies are designed and enacted in varied policy domains *in practice*, so as to consider how economic value is produced and circulated in *specific empirical settings*. The chapters that follow explore the production of markets and prices, as well as more general economic framings of value, in domains where value(s) were previously framed in non-economic terms. In exploring the emergence of newly economic values in a variety of case studies, the authors thereby also attend to the calculation and quantification of value, legitimacy and care, as well as to the implications of this quantification of value on human and beyond-human worlds. In conclusion, we consider some of the limits, hazards and consequences of the intensifications of economisation processes illustrated in our case research.

Context

This book builds on research conducted through a project designed in response to a call from the UK's Leverhulme Trust for research focusing on 'value'. It brings together a range of case studies carried out by PhD students, post-doctoral researchers and senior academics through the Leverhulme Centre for the Study of Value (www.thestudyofvalue.org). It also combines shared conceptual work that was carried out to simultaneously build a research protocol for case research, and to create alignment (where possible) in understanding the overlaps and differences exhibited by the Centre's diverse empirical research.

Our starting point is that at the heart of human life and social change is a profound uncertainty around the frontiers of current economic valuation practices. Neoclassical economics tends to pose as a complete and unified system of thought and pragmatics, but this posturing is, we feel, inaccurate and ultimately unhelpful in approaching the multiple socio-ecological challenges of the contemporary moment.

One often-observed problem is that there are frontiers to this utilitarian valuation system beyond which, in classical economic terms, lie 'externalities' rendered effectively value*less*. The composition of the 'valueless' changes historically according to moral, social and economic forces, but there is little explicit research on how social change and social struggles affect valuation frontiers, and vice versa (although on 'value struggles' see De Angelis 2007). Humanitarian and development efforts, for example, have reduced (but not eliminated) allowable deaths from famine, poverty, climate change, disease and war. Despite the human rights discourse however, a value of life for all has not yet been reached. We see this daily in struggles over the value of life of/for immigrants, refugees and other categories of people effectively treated as 'extras' on the stage of global capital and geopolitical contestation (cf. Mbembe 2003; Banerjee 2008; Sassen 2010; Gržinic and Tatlić 2014). More recently, loss of life in Syria, Yemen and Afghanistan is normalising a state of war and restoring the 'other' as terrorist, a condition exacerbated by Donald Trump's accession to the US presidency and his push for government endorsed xenophobia. Indeed, intolerance appears to be on the rise in many parts of the globe, rebuilding the number of abjected persons who are not deemed to be valuable. Development practice has instead generated various proxies such as 'quality-adjusted life years' and 'value for money' expenditure, which assume from the start that not every potential beneficiary of aid is equal (as liberal theory would suggest) or saveable.

Simultaneously, the efforts of social movements concerned with environmental and climate justice have brought more of the beyond-human world into economic calculations of value. An idea that has increasingly gained traction is that 'externalities' that become degraded because they are priced at zero by current economic structures will only be valued if incorporated with higher values through pricing mechanisms. As consumption of planetary resources is argued to approach finite limits (cf. Rockström et al. 2009; UNEP 2011), attempts to insist morally on intrinsic values of nature and other species have become set against work that quantitatively calculates, economises, marketises and sometimes financialises these natures (Robertson 2006; Pawliczek and Sullivan 2011; Sullivan 2013). This is the case, for example, in market arrangements for carbon offsetting and other 'environmental services', for marketised units of some species through species banking in the US, and for some habitats via REDD+[1] and through biodiversity offsetting instruments. Paradoxically, however, and against the avowed intent of those designing and implementing new calculative valuation practices, the quantitatively valued entities which emerge often appear to lose their earlier intrinsic value, emerging as more substitutable – and therefore more disposable – than ever.

Indeed, 'externalities' conventionally accorded zero-value in relation to market prices can offer new potential for capture and accumulation as they become calculated economically and priced. As such, newly calculated economic values for previously unpriced externalities can appear to become part of what anthropologist Anna Tsing (2015) describes as the relentless capture that continually engulfs the edges of formalised exchange relations, such that evermore entities and relations become co-opted to capitalism's ethos of monetary and profit-oriented value-making (see discussion in Sullivan 2018). This concern with how values are transformed so as to serve the interests of a particular political economic system echoes a Marxian sense of 'primitive accumulation' (Marx 1974[1867], 668) as the ontological, as opposed to historical, condition of capitalist production and value-making (De Angelis 2001; also Luxemburg 2003[1913]); Harvey 2010). A concern is that such value transformations work against possibilities for refraction of the mode of valuation that only sees 'that which is capital and that which might become capital' (Williams 2011, 30).

Such paradoxes of value and valuation are our focus in this book. The chapters in this volume explore a number of separate experiments in the articulation and calculation of value in the interconnected fields of policy and governance regarding development, environment and conservation. Most of the chapters engage with the development and application of specific 'calculative devices' (Callon 2007), i.e. new techniques and tools of quantification either working alongside traditional balance-sheet cost accounting at the firm level, or being used to influence social outcomes by public and private actors. Following Callon and colleagues (e.g. Çalışkan and Callon 2009, 2010; Callon and Muniesa 2005; Callon 2006), we argue that calculative devices are best understood in the context of the *assemblage* through which they are enacted, including the institutional setting and overall discursive framing and ideological representation of those involved. Calculative devices, institutional assemblages and discursive framing together form the three nodes of our research process for each empirical setting, as described in more detail in Chapter 2. Each of the ensuing chapters thus engages to varying degrees with these different nodes in the making of values that have come to matter in our cases. In working through empirical and textual detail in each of our cases our focus is to understand how a valuation process proceeds, and how, as a consequence of this process, a newly calculated and valued entity may be made and begin to act in the world.

In our emphasis on how valuation processes are enacted and their outcomes, we draw on the performative economics tradition with its focus on understanding how 'the economic' is made or 'performed', rather than simply describing a pre-existing economic reality (Çalışkan and Callon 2009, 2010; Callon 2007; MacKenzie 2006). In contrast with classical economic theory, which depicts 'the economy' as if it is a realm autonomous from human agency through the workings of an 'invisible hand' (e.g. Smith 2010[1759]; for critique see Polanyi 2001[1944]), performative economics theorises and explores the actions, knowledges, institutions and calculative technologies through which markets and marketed entities are created and maintained. As we unpack in the next chapter, Callon's concept of the 'sociotechnical *agencement*' or arrangement (STA) as a configuration of people, institutions and technologies that makes and performs markets, can assist with identifying how economic value is made in practice (Callon 2007;

Çalışkan and Callon 2010). The calculative devices within an STA effectively price and value entities for exchange, as demonstrated in studies on, for example, sulphur, greenhouse gas emissions, and fish (see Ellerman et al. 2000; MacKenzie 2009; Holm and Neilsen 2007), but the ways in which they do this are often infused with uncertainty, indeterminacy and contingency. This performative approach encourages observation of how economic performances take shape in practice, and thereby sheds light on the calculative and other machinations underscoring the exchange values that become visible in variously marketised structures of action and decision-making.

Our emphasis on how economic value is made thus assists with making visible some of the practices of assemblage that bring together multiple actors, materials, organisations, institutions, calculative devices, etc., that otherwise are mystified in the appearance and exchange of the commodity as an economically valued entity. The Marxian labour theory of value (LToV) similarly unmasks the fetish of value residing in the objectified commodity, by drawing attention to the labour and life captured or alienated in the making of commodities for generalised exchange (discussed further in Chapter 2). At the same time, Marxist value theory can take us only so far with regard to understanding how exactly new economised values are made in today's policy domains of development, environment and conservation. Analytical approaches are needed that both recognise the immeasurable contribution of Marx's critique of political economy and extend this contribution so as to connect with the technological and organisational specificities of the contemporary moment. We thereby read and apply performative economics approaches as disassembling and thereby clarifying the LToV by exploring the ways that multiple calculative devices, institutional practices, structuring value discourses and actions combine in practice to make new economic values from previously valueless domains. Our perspective, however, is that these performative elements exist within, and tend to shore up, the capital–labour relation so potently illuminated by the LToV.

On this basis the empirical cases that follow are more or less inspired by Callon's work on STAs, whilst also attending fully to the specificities or 'lumpiness' of the socio-ecological contexts they express in terms of power, race, inequality and social norms (cf. Collier and Ong 2008). All of the chapters use qualitative methods to capture the contextual detail and complexity at key sites and nodes of interaction, where policy-makers, institutions and calculative devices and technologies come together in valuation processes. These are fast-paced policy arenas and involve contemporary and ongoing framings of value within new markets and social spaces, demanding methodologies that draw in varying combinations on ethnographic closeness, expert interviews, actor network analysis, discourse analysis, institutional analysis, participant observation, semi-structured interviews, surveying, quantitative analysis and modelling. In many of the chapters, semi-structured interviews feature as a means to investigate the institutional settings, the actors and the discourses that are making the creation of new valued entities possible. In the remainder of this introductory chapter we briefly outline the key domains and case studies that form the basis of the chapters that follow.

Domains of value

The chapters in this book move through a series of interconnected, loosely boundaried 'domains of value' in contemporary policy arenas: development, environment and conservation. A domain, for the purposes of this book, is broadly conceived as a self-referencing space or conceptual terrain where participants have some collective convergence in the purposes, behaviours and activities that they are sharing. For example, the development domain encompasses a global set of institutions sharing a mission to 'do development' loosely conceived as a set of established practices and policies cohered in soft law internationally. People are employed in the domain, fiscal resources and finance are dedicated to it, and participants therein are expected or supposed to benefit from socio-economic and wellbeing gains. Similarly, in the environmental domain, there are also a set of institutions, persons and participants sharing a mission and purpose to improve the health of the beyond-human realm. The context of the first domain is society and the human, and of the second, other animals and the living biosphere. Conservation is closely related, but more specifically dedicated to ideas of preservation, restoration and ecological restitution and stewardship, often historically imbued with ideas of preserving wilderness. Moving through these interconnected domains of value, we proceed by summarising the different case studies making up the remainder of the volume.

The development domain

Historically there have been people who are not valued enough to be formally counted even when they die, as in the immeasurable loss of life in the colonial Congo, in the Chinese famine of 1961, or in the complexities of international intervention in the ongoing war in Syria. Lost lives have a particular political economy context, as the death of one person can improve economic opportunities for others, but the contemporary context of expendable people is not well understood. In development practice value calculations outside the price system are often used, since public actors must make decisions about how resources are to be distributed. But the process of quantification fails to respond to some of the biggest challenges of government: namely the morality that must underlie the principles and embedded assumptions that determine worth and value. Thus when desk officers at donor organisations decide that programming directed at helping orphaned and vulnerable children is better 'value for money' – and therefore a priority for expenditure – over programming directed at assisting people with disabilities or the elderly, they are using a complex calculation which may attach value to both the expected economic productivity of each group and a moral judgement over who is most deserving. The basis of such decisions is rarely made publicly explicit, but they result in inequalities and inequities in care alongside some of the improvements in performance that the planners had intended. Despite many studies that take a cost–benefit approach to assessing the value of development programming, there is little research that directly addresses the problem of how value is produced, and upon what basis.

Development effect, however, is increasingly quantified so as to service demands for 'evidence based policy', 'payment by results' and 'value for money'. To achieve these outcomes donors widely employ impact measurement technologies which perform quantified representations of how much of the 'good' thing desired by a public expenditure is actually being produced by a particular intervention. Impact measurement, however, often relies on classification and coding systems which invest numeric rankings with meaning in complex situations that resist the commensurability suggested by such numeric values. The role of impact measurement is often to convince others of the value of a policy intervention, rather than to prove impact in a scientifically robust way, such that the performance of value in development programming often suggests calculative worth beyond the actual capabilities of the measurement technology in question (Bracking 2012).

We begin this part with Chapter 3, in which Aurora Fredriksen explores the top-down work of valuation in the international development sector. In recent decades, international development agencies have become increasingly focused on demonstrating the effectiveness and impact of their work through practices of (e)valuation and assessment at the level of individual projects or programmes. Fredriksen examines one iteration of such 'aid effectiveness' accounting, namely 'value for money' (VFM), which must now be demonstrated for each project or programme funded by the UK government's DfID. Drawing on insights from assemblage theory and the performative economics literature, this chapter traces how VFM is variously calculated in practice and explores two related framings that circumscribe the possibilities of these VFM calculations. The first framing equates robust evidence with quantitative evidence. The second figures a linear relationship between development inputs and results. The chapter shows how the calculations of VFM within these framings performatively express the value of international development as a quantified, economic relationship between development inputs and results. The analysis is situated within the particular socio-political context in which this performance of VFM by DfID has been mobilised to defend development aid spending from its critics and considers the wider relevance of this dynamic.

Moving from a top-down to a bottom-up view of value in development, in Chapter 4 Fortunate Machingura focuses on ill-being and vulnerability, drawing on insights from the growing literature on surplus population, necropolitics, biopolitics and the expendability of human life (Foucault 1977, 2007; Mbembe 2003; Banerjee 2008; Murray Li 2007; Redfield 2012; Neves and du Toit 2013). In this chapter, Machingura explores the question of how societies implicitly and explicitly perceive and set the value(s) of different human lives through the discourses and calculative tools of public health policy and practice. In particular, the chapter identifies local and vernacular epistemologies and problematisations of value in relation to people living with HIV/AIDS (PLWHA) in Zimbabwe. The chapter identifies the calculative tools and rationalities used by government officials and public health experts in the valuation of human life and shows how the lives of people from certain social, cultural and economic backgrounds are thusly devalued. The chapter argues against the systematically 'allowable deaths' of these devalued populations, and for more transparent, just and equitable valuations of human lives in public health policy and provision. The chapter further demonstrates how PLWHA in Zimbabwe are not passive receptors of the (lack of) value bestowed upon them by these health systems, but instead resist this devaluation with a counter-narrative of their value in which they emerge not as expendable victims but as victors in an indefatigable population.

In Chapter 5, the final chapter on development value, Sarah Bracking and Aurora Fredriksen trace the competing financial and social valuations at play in the developmental worth of large-scale infrastructure projects, as well as the competing discourses used to justify them. This chapter starts with a case study of the 'Back of Port' expansion in South Durban and the competing logics of value articulated both in support of, and opposition to, a major port expansion and development project. The financial and economic logic of creating a fixed built asset to generate derivative income streams for investors, cheaper costs for shipping companies and expanded wharfs for the oil and natural resource export industries, is here juxtaposed to community views of the worth of the homes, social assets and heritage sites which would be displaced to make way for the port expansion. The case study is an example of competing orders of worth (Boltanski and Thévenot 2006): one abstract and derived from formal calculations of economic growth and multiplier effects; the other social and developmental and, in this case, largely absent from official debate. This chapter problematises any simple relationship between large-scale infrastructure and developmental benefit, arguing that the benefits to investors are the only guaranteed return in this project design. The valuation of the proposed port expansion is shown to prioritise profits at the international, rather than the national, scale, thereby shifting the benefits of infrastructure in favour of the (further) privatisation of profits offshore, while socialising the costs onshore. The chapter further argues that conventional development economics has difficulty grasping the value of such large-scale infrastructure projects in social terms, primarily because it occludes the issue of scale. Assuming a national frame of 'much-needed' Foreign Direct Investment (FDI) leads to a failure to account for the future financial, social and ecological costs of infrastructure builds.

The conservation domain

Environmental management for conservation is currently animated by attempts to make legible the value of beyond-human nature in cost–benefit decisions regarding economic development. Policy effort and funding are being directed towards creating calculative frameworks for 'valuing nature' that are global in reach, such that environmental externalities under conventional accounting practices can be clarified in terms of equivalent and apparently commensurable monetary representations (see, for example, the influential United Nations Environment Programme 'UNEP' initiative on The Economics of Ecosystems and Biodiversity or 'TEEB'). These efforts are permitting natures-beyond-the-human (cf. Kohn 2013) to be both conceptualised as, and aligned with, financial measurement, at the same time as facilitating the emergence of new marketised exchanges in these representations (Sullivan 2013, 2014, 2017a, 2017b). The offsetting and payments model for species conservation is rapidly accelerating as a core market-based method for valuing and conserving biodiversity involving varying combinations of public and private sources of finance. Although halting and beset by various sources of friction, these marketised and market-like approaches have significant implications in terms of both the conceptual disaggregation of species from the ecosystemic fabric in which they are embedded, and for the foreclosure of non-marketised motivations for valuing non-human nature. In this part two cognate case studies are included wherein natures-beyond-the-human are at the frontiers of monetary valuation practices. Through these practices natures are being configured as calculated and variously marketised units that can be incorporated into conventional neoliberal value orderings by means of formal property arrangements and particular efficiency and rationality assumptions.

In Chapter 6, Sian Sullivan highlights the value assumptions underscoring the recent and contested emergence of debt-based financing for the conservation of 'standing natures' *in situ*. In the aftermath of the United Nations Framework Convention on Climate Change (UNFCCC) Paris climate agreement of late 2015, there has been a noticeable proliferation of policy publications and reports espousing the benefits of leveraging debt-based finance and impact investing for the conservation of so-called natural capitals. The chapter examines how new measures of the health and improvement of natures *in situ* are becoming a focus for financial concerns in 'impact investing'. It introduces the concept and practice of impact investing and traces developments in the field of impact investing for the conservation, restoration and rehabilitation of terrestrial ecosystems. The chapter explores the innovations we are seeing today that seek to create conserved natures *in situ* as an asset class for impact investing, and sets them within a broader turn towards impact investing in social and development contexts. In doing so the chapter identifies the value-making mechanisms proposed and in place for translating projected natural capital quantities into financially leverageable forms of value.

Moving from the global to the local, in Chapter 7, Louise Emily Carver and Sian Sullivan analyse the applied practices that are creating numerical equivalences between ecological impact and associated offset sites so as to standardise measurable conservation outcomes, or 'yield', in the new conservation technology of biodiversity offsetting (BDO). The relative youth of biodiversity offsetting in application means that little is known either about how biodiversity valuations and offset contracts between impact and receptor sites are agreed in practice through application of biodiversity offsetting metrics, or the long-term implications for conservation outcomes. The chapter responds to the first of these gaps in particular, by demonstrating how biodiversity yields were calculated and negotiated for a specific offset contract, set within a broader government-led pilot study into biodiversity offsets in England. Empirical observations for how calculations of biodiversity value were enacted and applied in the development of a specific offset arrangement were made through repeat site visits, participant observations and semi-structured interviews with key actors participating over the 24 months of a biodiversity offsetting pilot study initiated by the UK government's Department for Environment, Food and Rural Affairs (DEFRA). In addition to 46 transcribed interviews subjected to textual analysis using a qualitative data analysis software program, biodiversity offsetting planning documents formed the basis for detailed analysis of this process in England. The case study presented here demonstrates how the DEFRA metric was applied in practice so as to create biodiversity value at both development and receptor sites and offers three central observations. First, that the standardising assessment techniques and metrics central to the biodiversity valuation process are being amended in creative ways as users adapt inputs to metric calculations to balance and negotiate conflicting requirements. Second, that the case history provides empirical material to illustrate the working in practice of commensuration processes that arise as different habitats are made equivalent through the application of BDO metrics. Finally that BDO is becoming further aligned with commercial agricultural productivity agendas that emphasise efficiencies and scale of production through concepts of 'yield'.

The environmental domain

International governance agreements under the auspices of the United Nations have, since the 1970s, drawn increasing attention to the environmental effects of economic development. In this part we consider two key dimensions of environmental governance: climate change mitigation and water management. Climate change futures is a rapidly growing area of research, but one in which natural and social scientists have encountered some difficulty in connecting their positivist and normative fields. Our research here considers both mature carbon markets and the morality of carbon trading through a case study of the UNFCCC, CDM and private sector carbon traders, alongside a specific exploration of discourses guiding and contesting the new GCF. Chapters in this part map the institutional arrangements and lay normativity within the carbon trading industry globally (Chapter 9), and in relation to the CDM (Chapter 8), in order to account for the apparent disjoint between claims that carbon trading is significant in reducing emissions, and global data suggesting that empirically this is not the case. Similarly, the question of water valuation has split the water sector for more than a decade. In particular, arguments about the role of water pricing and water markets have confronted quite different criteria rooted in the 'rights, knowledge and experience of local people and communities' (Chapters 10, 11). Increasingly, there is widespread recognition that 'price' does not, or should not, be equated with 'value'.

In Chapter 8, Robert Watt offers an explanation for why carbon markets have declined in terms of a number of interconnected meanings of value. The once-booming trade in CDM carbon credits has crashed, with offset credits now commanding only very low exchange values. A primary driver for the decline in offset markets is the growing perception that offsets have limited use value, especially in recognition of the failure of many offsets to produce the intended environmental use value, namely overall reductions in CO_2 emissions. The failure is related to the complexity of the valuation processes involved in creating an offset commodity. These include challenges in relation to measuring emissions reductions through the postulation of hypothetical 'baselines' and problems of attribution of emissions reductions to the offset mechanism. Both involve navigating the complex concept of 'additionality'. The commercial interests and power of the actors involved in the valuation process – mainly project developers and auditors, but also regulators – encourages the registration of significant amounts of offsets which lack environmental integrity. Policy-makers regulating the use of offsets in the European Union's (EU) carbon market have belatedly come to realise the lack of use value represented by offsetting, and have taken steps to restrict the trade. However, other users of offsets, including policy-makers governing other domains, might nonetheless evaluate carbon offsetting as a normatively commendable practice, or cynically recognise the problems of offsets and seek to exploit them. Offsets can be framed discursively as valuable, even when they are not, giving rise to an economy of false appearances (Tsing 2005; Cavanagh and Benjaminsen 2014). The chapter concludes that an inability to imbue meaningful use value to the carbon offset commodity through the CDM valuation process has led to a crash in the exchange value of these commodities.

In Chapter 9 Jonas Amtoft Bruun explores value in the GCF. At the international climate meeting (COP15) in Copenhagen in 2009, the UN GCF was conceived with the ambition that it would channel the majority of future financial resources to assist developing countries in adapting to the impacts of climate change. The GCF was further mandated to catalyse finance for mitigation projects under various existing UNFCCC programmes and to attract investments from the private sector via a 'private sector facility'. A total of 130 Civil Society Organisations (CSOs) participated in the four-year design process that took place from 2011 to 2015, as representatives for civil society and those most affected by climate change. This chapter explores the processes by which the activities of the CSOs influenced the design and development of the GCF. It argues that two meta-discourses prevalent within the UNFCCC process, green governmentality and ecological modernisation, worked in conjunction to enforce an economic calculus in the GCF design, which undermined values presented by the CSOs on human rights and pro-poor climate finance captured in a third meta-discourse on civic environmentalism. The chapter contributes to the study of value by critically assessing how valuation technologies in global climate governance are designed and the way in which value calculations render some knowledge and services provided by CSOs relevant, while others are deemed discountable.

In Chapter 10, Philip Woodhouse and Mike Muller review the question of value as it applies to water and explore how a diversity of water values can be reconciled within decision-making processes. They identify two separate issues to be addressed. The first is to characterise water as a complex resource with a diversity of uses, broadly defined, and to consider the values that may be associated with water in its different uses. The second is to consider how societies can and do make decisions about water resources where different values systems intersect and overlap. In particular they consider how conflicts over water resources can be mediated and whether there are generic principles that can guide the development and application of appropriate approaches. They critically review approaches that rely on reducing all water uses to commensurability consistent with a single basis of value, either economic or ecological. They then outline an analytical values-based approach that seeks to identify and recognise more explicitly a plurality of values in water resources and their use. They argue that resolving competition between different water users is more a matter of explicitly political processes than one of application of technical economic or hydrological rules.

In Chapter 11 Rebecca Peters and Philip Woodhouse explore the case of South Africa's 1998 National Water Act (NWA), which sought to address allocation disparities through the adoption of decentralisation, participation and economic pricing strategies for water governance and management. They argue that economic value as articulated through a 'scarcity' discourse emphasising efficiency, productivity and pricing for water has taken precedence over social equity values underpinning redistributive principles of water allocation reform (cf. Turton 2000). Drawing on empirical research in the Inkomati Usuthu Water Management Area in Mpumalanga province, they consider how the values of water expressed in the 1998 NWA overlap and differ across institutions, users and intended beneficiaries in practice. The chapter argues that the economic aspects of water value emphasised by dominant water users in this catchment, primarily commercial irrigated agriculture, have subverted the transformative potential of water reallocation following values based in notions of equitable allocation and citizens' rights to water. This has had the effect of entrenching historic patterns of inequitable access to water.

Conclusion

In this introductory chapter we have briefly framed the structure and rationale for this book. We hope our volume will inspire further comparative empirical research regarding the making of new economic value in varied contexts, as well as contribute to the theorisation of value and valuation processes in both the academy and beyond. Given that the work is both theoretical and empirical, the next chapter is reserved for a more detailed account of our conceptual map for the research, and the concepts which we decided to use in a common research protocol.

Note

1 REDD+ refers to Reducing Emissions from Deforestation and Forest Degradation, a mechanism developed by the UNFCCC to create a financial value for the carbon stored in forests in developing countries.

References

Adler, S. (producer) (2013) "Worth: How do you put a pricetag on nature?" *Radiolab* [audio podcast], season 13, episode 3. www.radiolab.org/story/worth/ (Accessed 20 May 2016).

Bakker, K. (2003) *An Uncooperative Commodity: Privatizing Water in England and Wales*. Oxford: Oxford University Press.

Banerjee, S.B. (2008) "Necrocapitalism" *Organization Studies*, 29: 1541–1563.

Boltanski, L. and Thévenot, L. (2006) *On Justification: Economies of Worth*. Princeton, NJ: Princeton University Press.

Bond, P. (2013) "Africa 'rising', South Africa lifting? Or the reverse?" *Daily Maverick* 6 February 2013, www.dailymaverick.co.za/opinionista/2013-02-06-africa-rising-south-africa-lifting-or-the-reverse/#.UYk7KMpdDW8 (Accessed 10 December 2017).

Bracking, S. (2012) "How do investors value environmental harm/care? Private equity funds, development finance institutions and the partial financialization of nature-based industries" *Development and Change*, 43(1): 271–293.

Çalışkan, K. and Callon, M. (2009) "Economization, Part 1: shifting attention from the economy towards processes of economization" *Economy and Society*, 38: 369–398.

Çalışkan, K. and Callon, M. (2010) "Economization, Part 2: a research programme for the study of markets" *Economy and Society*, 39(1): 1–32.

Callon, M. (2006) "What does it mean to say economics is performative?" *CSI Working Papers Series* 005. 2006. https://halshs.archives-ouvertes.fr/halshs-00091596/document (Accessed 7 July 2017).

Callon, M. (2007) "What does it mean to say that economics is performative?", pp. 311–357 in D. MacKenzie, F. Muneisa and L. Sui (eds) *Do Economists Make Markets? On the Perfomativity of Economics*. Princeton, NJ: Princeton University Press.

Callon, M. and Muniesa, F. (2005) "Economic markets as calculative collective devices" *Organization Studies*, 26: 1229–1250.

Cavanagh, C. and Benjaminsen, T.A. (2014) "Virtual nature, violent accumulation: the 'spectacular failure' of carbon offsetting at a Ugandan National Park" *Geoforum*, 56: 55–65.

Collier, S.J. and Ong, A. (2008) *Global Assemblages: Technology, Politics, and Ethics as Anthropological Problems*. Oxford: Wiley.

De Angelis, M. (2001) "Marx and primitive accumulation: the continuous character of capital's 'enclosures'" *The Commoner* 2, www.commoner.org.uk/02deangelis.pdf (Accessed 22 October 2017).

De Angelis, M. (2007) *The Beginning of History: Value Struggles and Global Capital*. London: Pluto.

Ellerman, A.D., Joskow, P.L., Schmalensee, R., Montero, J.P. and Bailey, E.M. (eds) (2000) *Markets for Clean Air: The U.S. Acid Rain Program*. New York: Cambridge University Press.

Foster, J. (ed.) (1997) *Valuing Nature? Ethics, Economics and the Environment*. London: Routledge.

Foucault, M. (1977) *Discipline and Punish: The Birth of the Prison*. New York: Pantheon Books.

Foucault, M. (2007) *Security, Territory, Population: Lectures at the Collège de France 1977–78*. Basingstoke: Palgrave Macmillan.

GCF (2017) *Green Climate Fund* [Website] https://www.greenclimate.fund.

Gržinic, M. and Tatlić, S. (2014) *Necropolitics, Racialization, and Global Capitalism: Historicization of Biopolitics and Forensics of Politics, Art, and Life*. New York: Lexington Books.

Guyer, Jane (2004) *Marginal Gains. Monetary Transactions in Atlantic Africa*. Chicago, IL: University of Chicago Press.

Harvey, D. (2010) *The Enigma of Capital and the Crises of Capitalism*. London: Profile Books.

Helm, D. (2015) *Natural Capital: Valuing the Planet*. London: Yale University Press.

Holm, P. and Nielsen, K.N. (2007) "Framing fish, making markets: the construction of individual transferrable quotas" *Sociological Review*, 5(2): 173–195.

Kjellberg, H., et al. (2013) "Valuation studies? Our collective two cents" *Valuation Studies*, 1(1): 11–30.

Kohn, E. (2013) *How Forests Think: Towards an Anthropology beyond the Human*. Oakland: University of California Press.

Lohmann, L. (2009) "Toward a different debate in environmental accounting: the cases of carbon and cost–benefit" *Account, Organizations and Society*, 34: 499–534.

Luxemburg, R. (2003[1913]) *The Accumulation of Capital* (trans. A. Schwarzschild). London: Routledge.

MacKenzie, D. (2006) *An Engine, Not a Camera: How Financial Models Shape Markets*. Cambridge, MA: MIT Press.

MacKenzie, D. (2009) "Making things the same: gases, emission rights and the politics of carbon markets" *Accounting, Organizations and Society*, 34(3–4): 440–455.

Martin, A., McGuire, S. and Sullivan, S. (2013) “Global environmental justice and biodiversity conservation” *Geographical Journal*, 179: 122–131.

Marx, K. (1974[1867]) *Capital*, vol. I (trans. S. Moore and E. Aveling). London: Lawrence and Wishart.

Mbembe, J-A. (2003) “Necropolitics” *Public Culture*, 15(1): 11–40.

Murray Li, T. (2007) *The Will To Improve: Governmentality, Development, and the Practice of Politics*. Durham, NC: Duke University Press.

Neves, D. and du Toit, A. (2013) “Rural livelihoods in South Africa: complexity, vulnerability and differentiation” *Journal of Agrarian Change*, 13(1): 93–115.

O’Neill, J. (2007) *Markets, Deliberation and Environment*. London: Routledge.

Pawliczek, J. and Sullivan, S. (2011) “Conservation and concealment in Species-Banking.com, USA: an analysis of neoliberal performance in the species offsetting service industry” *Environmental Conservation*, 38: 435–444.

Polanyi, K. (2001[1944]) *The Great Transformation: The Political and Economic Origins of Our Time*. Boston, MA: Beacon Press.

Poovey, M. (1998) *History of the Modern Fact: Problems of Knowledge in the Sciences of Wealth and Society*. Chicago, IL: Chicago University Press.

Porter, T. (1995) *Trust in Numbers: The Pursuit of Objectivity in Science and Public Life*. Princeton, NJ: Princeton University Press.

Power, Michael (1997) *The Audit Society: Rituals of Verification*. Oxford: Oxford University Press.

Power, M. (2003) “Auditing and the production of legitimacy” *Accounting, Organizations and Society*, 28(4): 379–394.

Redfield, P. (2012) “Bioexpectations: life technologies as humanitarian goods” *Public Culture*, 24(1): 157–184.

Robertson, M.M. (2006) “The nature that capital can see: science, state, and market in the commodification of ecosystem services” *Environment and Planning D: Society and Space*, 24(3): 367–387.

Rockström, J., et al. (2009) “Planetary boundaries: exploring the safe operating space for humanity” *Ecology and Society*, 14, art. 32, www.ecologyandsociety.org/vol14/iss2/art32/

Sandel, M. (2013) *What Money Can’t Buy: The Moral Limits of Markets*. London: Penguin.

Sassen, S. (2010) “A savage sorting of winners and losers: contemporary versions of primitive accumulation” *Globalizations*, 7: 23–50.

Smith, A. (2010[1759]) *The Theory of Moral Sentiments*. London: Penguin.

Sullivan, S. (2013) “Banking nature? The spectacular financialisation of environmental conservation” *Antipode*, 45: 198–217.

Sullivan, S. (2014) “The natural capital myth; or will accounting save the world? Preliminary thoughts on nature, finance and values” *LCSV Working Paper Series No. 3,* http://thestudyofvalue.org/wp-content/uploads/2013/11/WP3-Sullivan-2014-Natural-Capital-Myth.pdf (Accessed 7 July 2017).

Sullivan, S. (2017a) “On ‘natural capital’, ‘fairy-tales’ and ideology” *Development and Change,* 48: 397–423.

Sullivan, S. (2017b) “Making nature investable: from legibility to leverageability in fabricating ‘nature’ as ‘natural capital’” *Science and Technology Studies*. https://sciencetechnologystudies.journal.fi/forthcoming/article/58040/27301

Sullivan, S. (2018) “On possibilities for salvaged polyphonic ecologies in a ruined world” *Dialogues in Human Geography,* 8(1): 69–72.

Ten Kate, K. and Crowe, M. (2014) *Biodiversity offsets: policy options for governments. An input paper for the IUCN Technical Study Group on Biodiversity Offsets.* International Union for Conservation of Nature, Gland, Switzerland. URL: www.forest-trends.org/documents/files/doc_4777.pdf

Tsing, A.L. (2005) *Friction: An Ethnography of Global Connection.* Princeton, NJ: Princeton University Press.

Tsing, A.L. (2015) *The Mushroom at the End of the World: On the Possibility of Life in Capitalist Ruins.* Princeton, NJ: Princeton University Press.

Turton, A.R. (2000) "Precipitation, people, pipelines and power in southern Africa: towards a 'virtual water'-based political ecology discourse", pp. 132–153 in P. Stott and S. Sullivan (eds) *Political Ecology: Science, Myth and Power.* London: Edward Arnold.

UNEP (2011) *Towards a Green Economy: Pathways to Sustainable Development and Poverty Eradication. A Synthesis for Policy Makers*, available from www.ipu.org/splz-e/rio+20/rpt-unep.pdf (Accessed 7 January 2014).

Wilde, O. (2005[1892]) *Lady Windermere's Fan, A Play About a Good Woman.* London: Nick Hern Books.

Williams, E.C. (2011) *Combined and Uneven Apocalypse.* Ropley, Hants, UK: Zero Books.

2 Value(s) and valuation in development, conservation and environment

Sarah Bracking, Aurora Fredriksen, Sian Sullivan and Philip Woodhouse

The chapters in this collection explore ways in which new economic values are being created, assessed, assigned and understood in a variety of contexts in the domains of social development, conservation and environmental governance. These chapters variously interrogate valuation practices and the calculation and production of economic value through the interplay of moral and economic logics in a number of contexts. These encompass the calculation of value in development programming and aid decisions; government-supported markets for offsetting biodiversity loss and carbon emissions aimed at addressing the coupled global environmental crises of biodiversity loss and climate change; and contested efficiency criteria used to justify water allocation in Africa. Although covering a diverse set of policy domains, the chapters that follow are united in an empirical approach to value that focuses on 'how' questions. How are newly economic values being made, performed and represented in the fields of development, conservation and environment policy and practice? How are various calculative tools being developed and used to measure and represent new values? How do certain discourses come to frame these values, how are they arrived at, and how do they circulate and produce effects in policy and practice? And, finally, through what institutional assemblages are valuation practices promoted and new values articulated?

By studying how valuation processes proceed empirically, these collected chapters aim to understand the deficiencies and possibilities of contemporary economic valuation practices in political, social and environmental terms. We thereby focus on the production of 'value' through the creation of new 'value entities' that can be accounted for, costed and circulated in monetised and financialised forms, such as within a market in which they have a price. Each of the chapters draws on a common framework for studying valuation processes empirically: identifying the calculative devices, calculative technologies, institutional assemblages and discursive framings that go into the process of producing new 'value entities'. In so doing, we join a growing group of scholars interested in interrogating 'value' as something produced and performed through practices that can be traced empirically. This is not to say that values – once produced and performed – do not have real effects in the world. Indeed, the collected case studies in this volume show otherwise. Rather, this collection affirms that economic values do not pre-exist their measurement or articulation as explanatory entities in their own right, but instead are themselves phenomena in need of explanation.

As we recognise in the next section, we are not the first to assert this. It is important, however, to set this assertion against a background in which economic valuation has come to dominate political narratives of value over the past half-century. In so doing, we can more clearly locate the empirical processes through which this occurred. A number of elements can be identified that suggest an interlinking of economic and environmental thinking in the 1960s and 1970s in ways that were logical but that evolved in ways that were not necessarily foreseen, much less predetermined. A first element can be identified in the development of economic analysis of "social costs" (Coase 1960) and its instrumentalisation as cost–benefit analysis (CBA) as a means of determining whether the aggregate benefits of a proposed initiative outweighed the overall costs. A parallel concern with impacts mediated through the biophysical environment culminated with the notion of environmental impact assessment (EIA). These approaches were initially distinct in that CBA was concerned with effects that could be measured in economic terms and were thus reducible to a single positive or negative figure expressing the balance between the two. In contrast, environmental impact assessment made no claim to aggregate the effects it identified to a single measure, but emphasised plurality in multi-criteria assessment.

The development of environmentalism in the 1970s generated narratives that transformed this separation between the aggregating approach of economics and the pluralist assessment of incommensurable phenomena in EIA. In particular, narratives of a finite planet and fragile ecosystems characterised nature in terms of loss and scarcity. The "limits to growth" (Meadows et al. 1972) projected at a global scale were used to provide an ecological narrative of unsustainable human population growth to explain instances of human vulnerability, most notably in the case of famine following drought in Africa. During the 1980s one response was to construct a sustainability narrative in which ecological security required the elimination of social deprivation (WCED 1987). Another was to extend to the 'problem' of the environment the logic of neoclassical economics. In the latter, the degradation of nature and its ensuing 'scarcity' was a consequence of undervaluing nature: if the value of natural elements such as forests, wetlands, wildlife, etc. could be correctly revealed through prices, then scarcity would be reflected in higher prices and reduced demand (i.e. conservation). In this neoliberal 'counter-revolution' scarcity was removed in so far as the limits of environment could be economically defined, and progressively monetised and sometimes traded (Nelson 2014, 2015). While the privileging of market 'solutions' reflected the neoliberal political domination of the era, the plausibility of market valuation as a route to nature conservation hinged critically on the perception that it provided a means by which society would be made more aware of the value of what was being lost. So forests' economic valuation was extended from the price of timber to the calculation of equivalent financial benefits derived from non-timber forest products, such as medicinal, fibre and food plants harvested from the standing forest, often with the express aim of demonstrating the greater economic value of leaving timber uncut (e.g. Peters 1996).

In its early manifestations (cf. Pearce et al. 1989), therefore, environmental economics emphasised the importance of finding hidden economic value in order to construct valuations of nature that would counter the financial returns claimed by developers. Such arguments for the economic benefits from nature conservation only worked, of course, if these benefits were actually realised: a matter of 'use it or lose it'. Hence the importance of realising the economic benefits of African wildlife, for example, in terms of revenues from tourism to local communities to counter the foregone benefits of expanding cultivation and livestock grazing. The move from making visible economic values in nature to using such values in making decisions about 'trade-offs' between costs and benefits of alternative activities and ultimately using such values to trade and commodify elements of nature may appear inexorable. Yet the development of markets in environmental commodities, as with all markets, demonstrates the critical role of non-economic factors. As the following chapters show, the work entailed in establishing a value for something, making that value economic, and then making that economic value act like a commodity is extensive and the results are often unstable.

The critical role of non-economic factors may emerge as directly political, or as administrative fiat. For example, it may be a matter of blatant collusion in price-fixing, as in "wetland banking" (Robertson 2004), but more fundamentally the establishment of markets such as emission trading show an interplay between context-specific political expediency and a broader ideological commitment to promote market allocation as a preferred governance mechanism. As documented by Gorman and Soloman (2002) and Lane (2012), the development of emission trading in the US in the 1970s did not emerge from a policy blueprint, but from a need to reconcile contradictory political imperatives of compliance with air quality standards and enabling new industrial development. Local authorities in industrial areas with poor air quality sought to 'offset' emissions from new industry by requiring those responsible for new emissions to achieve reductions either in their existing operations or in those of neighbouring companies. Lane (2012) has observed that the translation of such local arrangements to a policy of emissions trading at a national level required a series of narratives that altered (or re-framed) perceptions of the problem of clean air. Specifically, the political problem of poor air quality which pitted local residents against industrial polluters was re-framed as a problem of aggregate pollution (i.e. tons of pollutants emitted per year) and the efficiency with which this could be reduced. This could only be achieved if mandatory and uniform air quality standards, and penalties for non-compliance such as carbon taxes, could be portrayed as 'inefficient'. That this was achieved was due to political lobbying by industry who sought to minimise the costs of compliance. The primary goal of improved air quality for residents was thus replaced by a goal of reduced cost of compliance for industry.

This brief reflection on the rapid and radical domination of valuation by economics over the past half-century bears out not only the importance of understanding how values and systems of valuation come about through political process, but also the contingency of such developments on particular historical circumstances. There is thus a historical context to this book in relation to the broader movement of political economy over the last 50 years or so, which has tended to a privileging of economic and then arguably financial imperatives in the organisation of human and non-human affairs. There are specific contexts, spaces and geographies in which valuation processes, which in this book are conceptualised within institutional assemblages, have been increasingly privileged. Moreover, economic values are created by multiple agents in circumstances where economics is not the only thing of concern, and yet economic criteria have grown to dominate many of these valuation processes despite the salience and resonance of other variables. Thus we have set out to explain how this dominance happens, in time and space, in our various case studies. We have also discovered that sometimes it does not, and we return to this alternative finding in the conclusion.

Value, values, valuation

The Oxford English Dictionary defines 'value' as "[w]orth or quality as measured by a standard of equivalence".[1] Beyond this basic definition, 'value' and 'values' have come to signify a multitude of diverse, often conflicting, meanings within different academic traditions. In the interest of space, we do not retrace all of these lineages here (for extensive discussions of value in the social sciences see e.g. Graeber 2001; Lamont 2012; Skeggs 2014; Stark 2009). We are also limited in space in relation to studying the specific trajectories of these considerations of value in each social science, through, for example, the seminal work of anthropologists and sociologists like Pierre Bourdieu (and exchangeable forms of capital), Stephen Gudeman, Jane Guyer, Keith Hart, Mary Douglas, Nicholas Thomas, Marilyn Strathern (in particular, [2000] *Audit Cultures*), Marshall Sahlins and David Mosse. For economists, value is a core concern, and beyond price theory and mainstream economics, where ideas of resource scarcity and distributional matters tend to be dominant, value often has a central place. For example, Marxist economics hinges on a central law of the Labour Theory of Value, where all circulating value is said to derive from workers' labour. Also, in the broad field of economics, heterodox theorists are invariably encouraged to find value beyond the mere price of things to provide an epistemology for sustainability and human-centred development in such concepts as impact investing and the triple bottom line. Because value is such a core concern in social science we would struggle to cover every contribution in a book this size, while still attempting to offer something new. So instead we offer a more selective account of the theories of value we have drawn on, for reasons of both expedience and relevance, in our own interrogations of value in policy and practice in development, environment and conservation domains. While offering apologies in advance to writers who feel excluded, and for any scholarly deficiencies that might arise from this more selective account, we do propose that what is offered here has a pattern of literature preferences that we feel best suit a study of value at this particular point in time. We will return to this body of work again in the conclusion.

Our collective studies on the current rise of economic valuations draw from a long lineage of work on relationships between economic and governmental values and rational calculation: Weber (1930[1905]) persuasively showed how a "calculating rationality" was associated with the emergence and consolidation of capitalism; Foucault (1991[1975]), 2007) argued that calculative frames and devices that abstract, measure and rationalise bodies and nature are key to the mode of governmentality of the modern state; Deleuze and Guattari (1987[1980]) similarly observed that productive 'striations' of time, space, bodies, nature, etc. are associated with the consolidated apparatus of what they term "State Science". Graeber's *Toward an Anthropological Theory of Value* (2001) is also seminal here. He identifies three "large streams of thought that converge" in the meaning of 'value', these being:

> 'values' in the **sociological sense**: conceptions of what is ultimately good, proper, or desirable in human life; 'value' in the **economic sense**: the degree to which objects are desired, particularly, as measured by how much others are willing to give up to get them; (and) 'value' in the **linguistic sense**, which goes back to the structural linguistics of Ferdinand de Saussure (1966), and might be most simply glossed as 'meaningful difference'.
>
> (Graeber 2001, 1–2)

In our explorations of how a particular thing might become subject to calculative processes, and then become subsequently valued differently through that engagement, we seek to work with how these differing conceptions of value perform, relate, condition, reinforce and contradict each other in our different empirical cases.

Before we turn more fully to the epistemology and method guiding the research presented in the remainder of this volume, however, we engage further with some key prior framings of how value is made. It is of course impossible to engage with concerns regarding the making and performance of economic value without also engaging with the Labour Theory of Value (LToV) as presented in the writings of Karl Marx. While we depart somewhat from Marxist understandings of value as a "static category revolving around capital and labour" (Dussauge et al. 2015, 17), we are alert to the influence of Marx's positioning of value as composed of, and derived from, social relations obscured by the appearance of value as something inherent to commodities. The LToV is potent precisely because it unmasks the fetish of value residing in the objectified commodity set "opposite living labour as an alien power" (Marx 1993[1857–1858], 454). In doing so the LToV draws attention to dimensions of activity, materiality and life that become captured through processes of commodity-making but then hidden from view in the appearance of the commodity. The LToV thus points to both the domains of life and activity from which capitalist (exchange) value is alienated (Taussig 2010[1980]; Holloway 2015), and to the unequal accumulations of surplus value – manifest as both profit and rent – driving the engine of commoditised production requiring this alienation (cf. Luxemburg 2003[1913]).

While some scholars tend to juxtapose new theory as inevitably at odds with classical Marxism, we instead affirm that new empirical work on valuation, including work presented in this volume, builds on and adds to Marxist value theory in important ways (also see Henderson 2013). For example, this work breaks down the fixity of certain economic entities and dyads – such as labour and capital, state and market, private and public – borrowed in Marxism from the classical political economists of the 18th century liberal tradition, with all the associated heuristic characteristics of Modernism. To 18th and 19th century intellectuals, proving neatness, fixity, laws and science was part of the modernist project, and the production of a law around a singular theory of value was Marx's contribution, among others, to that work. Beyond its proclamation as a law, however, Marx's labour theory of value is flawed because its units can only be inferred: the units of socially necessary abstract labour time (SNALTs) required to produce a commodity cannot be exactly measured or known. This immeasurability is shared with its counterpart the *util*, i.e. the base unit of utilitarian economics that hypothetically measures satisfaction (Bichler and Nitzan 2015). Instead, as Bichler and Nitzan summarise, they can only to be known by their effects (2015).

Embracing complexity, relationality or contingency instead requires rejecting the fixed laws of traditional political economy, an observation particularly salient given the complexity of 21st century economic realities (Mandelbrot 2008). As Bichler and Nitzan note of mainstream economics:

> You cannot build an entire social cosmology [economics] on the assumptions of individual rationality, equilibrium and perfect markets – and then blame the failures of this cosmology on irrationality, disequilibrium and imperfections. In science, these excuses and blame-shifting are tantamount to self-refutation.
>
> (Bichler and Nitzan 2015, 67)

Similarly, for the LToV to work as a concept we must accept that all commodity value comes from the application of human labour (in combination with the (re)generative productivity of beyond-human natures

(Gareau 2005)), and that socially necessary abstract labour time emerges magically in an accumulation process that looks familiar to its sibling, the perfect competitive market: only in competitive markets could the generalised and averaged value of a unit of SNALT emerge. The limitations of such a theory have been made manifest in a generation of work from feminist and post-colonial scholars, as well as others, which show the very particular roles of gender, race, power and geography, in structuring markets in particular non-economic and non-competitive ways. Thus, just as we understand real-world markets as constituted by more diverse social processes than allowed by their abstract ideal, so also with value and valuation. Foudationally, since neither utils nor SNALTs, or the assumptions behind their conceptualisation, lead us to know more about value and valuation, we have chosen instead an epistemology that breaks the link to a foundational LToV (or util) that cannot be measured or refuted (on which more below). This approach enables exploratory empirical possibilities (as demonstrated in the chapters that follow), as well as offering a renewed recognition of the interconnections between things, people and other-than-human natures in their co-contribution to that which is and becomes valued.

While we draw from Marx, then, the idea of value as a social relationship, we also draw from the pragmatist tradition a sense that value is not something that exists 'out there' as an entity unto itself, but is rather something that can only be known from the practices through which it is produced or performed. In the pragmatist tradition, John Dewey argued that, strictly speaking, there is no such thing as value but only things possessing "the unique, the experienced, but undefinable, quality of value" and that, moreover:

> [v]alues in the plural, or value in the singular, is merely a convenient abbreviation for an object, event, situation, res, possessing the quality. Calling the thing a value is like calling the ball struck in baseball a hit or a foul.
>
> (Dewey, 1923, 617; cited in Muniesa 2012, 25; see also Dewey 1939)

In this perspective, then, the only way that 'value' can be understood is by attending to the practice (or action) of valuation.

The pragmatist tradition of focusing on actual practices resonates with (most) actor network theory (ANT), after-ANT, and assemblage theory analyses of value. For Michel Callon and his followers, for example, value arises through ongoing performance, or "performation", by dynamic socio-technical assemblages or agencements (STAs). Boltanski and Thévenot (2006[1991]) similarly see value (or worth) as arising through ongoing and dynamic (re)enactments requiring continual negotiations and innovations in everyday life. Michèle Lamont (2012) defines valuation as the action of "giving worth or value" through practices of categorisation – i.e., classification, commensuration, signalling and standardisation – as well as of the social legitimation of the categories that thereby arise – through contestation and negotiation, diffusion, stabilisation, ritualisation, consecration and institutionalisation. Lamont (2012) and other sociological researchers share kinship with this pragmatist emphasis on practices of valuation although follow Bourdieu (e.g. 1993) in approaching the study of valuation (and evaluation) as the study of the cultural practices associated with assigning relative worth within different 'fields', for example economy, politics, law, art or education (also Boltanski and Thévenot 2006[1991]). Attending to the social and cultural negotiations and contestations arising over questions of what constitutes appropriate criteria for valuation and who can be a legitimate judge of value in various fields, are thus central to this approach (Bourdieu 1993; Lamont 2012; Boltanski and Thévenot 2006[1991]).

There are resonances of this performative orientation to value in the geography literature associated with notions of 'political ecology'. More specifically, critical reassessments of the relationships between human society and non-human 'nature' have emphasised both the importance of human subjectivity in the 'construction of nature' (Smith 1984; Cronon 1991) as well as the significance of materialities in accommodating or frustrating the plans of human society (Castree 2002, 2003; Bakker and Bridge 2006). The discussion of the relative agency of non-human nature has sought to avoid an environmental determinism on the one hand and a subordination of nature to human subjectivity on the other, leading to notions of "hybridisation" and "quasi-objects" (Swyngedouw 1996, 1999). For our purposes here, the significance of these discussions is that they locate the qualities – and 'value' – of elements of nature as arising from their relations (or geometries of power) with human society in historically and spatially specific contexts. This understanding of context-specificity of valuation is congruent with that of O'Neill (2007) who asserts that people often value habitats not in ahistorical terms of itemised ecosystem services but through historically specific associations or relationships.These include, for example, childhood memories or a sense of landscapes as manifestations of the lives and identities of particular communities in farming, tin-mining, quarrying or whatever.

Although less concerned with theories of 'culture' and more with attending to empirical parameters, in the American pragmatist tradition 'valuation' also emphasises the actions associated with assigning value/s. John Dewey, a central early thinker in the American pragmatist tradition, thus argued for a focus on valuation as action, process or practice, rather than on value as something with either a real or ideal basis (Dewey 1915a, 1915b, 1923,1939). Such a move, Dewey suggested, can effectively intervene in the dualism of an 'idealistic–realistic' divide, not by choosing one side or the other, but by way of providing an alternative means of understanding, or 'flank movement' (Muneisa 2012), which decentres the idealistic–realistic duality. More specifically, Dewey argued that:

> The situation in which judgement of value is required is not mental, much less fanciful. It is existential, but it exists as something whose

> good or value resides (first) in something to be attained in action and (secondly) whose value both as an idea and as existence depends upon judgement on what to do. Value is 'objective,' but it is such in an active or practical situation, not apart from it.
>
> (Dewey 1915a, 516; quoted in Muneisa 2012, 26)

The pragmatist approach thus locates both the reality and ideation of value in empirically traceable action, suggesting that the two are not entirely separable in practice. In modernity, traceable acts of valuation and the ideas by which they are infused are frequently associated with innovations in measurement, standardisation and numerical attributions of monetary worth. Weber (1930[1905]; also Foucault 1991[1975]; Federici 2004) famously emphasised this impetus of 'calculative rationality', arguing for its material effects in terms of shaping society and economy. In this view, calculative rationality and the apparatus of calculation with which it is entwined are important in four ways in creating values that come to matter: first, by acting historically to generate consistent patterns in the distribution and organisation of wealth; second, by extending an emphasis on particular forms of value and value-making activities/institutions (namely money as measure of all value); third, by amplifying particular subjectivities, including that we are self-interested rational accumulators of particular forms of value, and that we have no kinship with non-human nature; and finally by fostering a spectrum of exclusions, inequities and environmental degradations. Our work builds on this tradition by examining the specific valuation technologies and processes which contribute to these causal and consequential effects of valuation. Studying calculative rationality enables us to identify different calculative devices and technologies, notional values and value framings and the ways in which these are used to incorporate ever more things into socially articulated markets and spaces, reconstructing them as new 'valued entities' with a numerical or calculated worth.

With these diverse, but interrelated understandings of value in mind, our collective project in this volume is to interrogate the ways in which value in the contemporary moment is produced, enacted and circulated in the world. These performative approaches amplify the insights of the LToV by disassembling, and thereby clarifying, the world-making actions of multiple calculative devices, institutional practices and structuring value discourses that in combination make new economically valued units that often can also be traded as commodities. The ethnographic and data-rich analyses that such approaches encourage illuminate not only the calculative and other machinations underscoring the exchange values that become visible in variously marketised structures of action and decision-making, but also alternative rationalities that may contest and creatively subvert them (Taussig 2010[1980]). As such, an emphasis on how new economic value is made can assist with making visible some of the practices of assemblage that bring together multiplicitous actors, materials, organisations, institutions and calculative devices around particular value(d) entities, as well as some of the practices of contestation through which these entities might be unmade.

Research protocol for creating values that matter

Having outlined the above key approaches to the creation of values that matter in the contemporary moment we now proceed to delineate our specific protocol for researching the creation of values that matter in our cognate case studies in development, conservation and environment domains of value. As indicated briefly in Chapter 1, at the beginning of our four-year programme of research, we engaged in a collaborative process to design a research methodology to guide our case-study research without dictating how research should be conducted in each case context. All of the chapters that follow draw on this protocol to varying degrees. It involves the identification and consideration of three key interacting elements of valuation processes in each case, namely calculative technologies and devices, institutional assemblages and practices of assembling, and discursive framing (also see Corson et al. 2013; Sullivan forthcoming).

1. Calculative technologies and devices

We understand a calculative *device* to be a specific package or technique, such as a scorecard or equation, and a calculative *technology* to include the purpose and context in which a calculative device is used. Calculative devices are the statistical packages, software programs, equations, pro formas, audit technologies and so forth that can be used to measure a thing or person in some attribute. For example, the International Finance Corporation (IFC) employs a calculative device, the Development Outcome Tracking System (DOTS), in order to measure the development impact of its investments in private sector development. Here the DOTS scorecard (the indicators and their weightings) is the calculative device; while what is being done, i.e. development impact assessment, is the calculative technology. This impact assessment is done in a particular way, within specific institutions (normally by consultants contracted to the IFC), which together make up an institutional assemblage for performance standards and safeguarding within the private sector development industry. This assemblage promotes particular understandings, or discursive framings, of the process, which in this case can be summarised as deriving from a world view in which economic growth is valued, and consultants' expertise is a professional and technical attribute understood as guaranteeing 'development' value. In emphasising this constitutive combination of calculative device, calculative technology, institutional assemblage and discursive framing, our cross-case research protocol provides a means to empirically explore, in this example, development impact valuation (see Chapter 3 this volume; also Bracking and Ganho 2011). What emerges at the end is a new, partially abstract 'valued entity', for example 'private sector development'.

The case-study areas we explore in this volume thus assist with identification and investigation of the 'real world' effects on human and environmental wellbeing of contemporary and expansionary calculative and market-oriented technologies as they 'render technical' previously uncalculated domains of being (cf. Li 2007a, 263). All our case studies contain a valuation process embedded within a 'calculative technology' which probably incorporates one or more specific calculative devices that act to create new numeric value entities in pecuniary and non-pecuniary terms. A pattern emerges across our cases in that many of the valuation processes we are studying act to replace or obscure non-pecuniary prior value with values in terms of price (for new markets cases, see, for example, Chapter 7 by Carver and Sullivan in this volume) or costs (as in the cost–benefit models developed around HIV/AIDS treatments and allowable death in Machingura's case research and for costing climate change in Watt's research (Chapters 4 and 8, respectively, in this volume)). Each of our case studies focuses on the production and performance of 'value' by mapping the valuation process in which new 'value entities' are made, and how these can be accounted for, costed and sometimes also circulated as commodities in monetised and financialised forms.

In our volume new value entities include 'allowable death', carbon credits, biodiversity offsets, water allowances and development impact scores. The important contribution of the concept of 'calculative technologies' is that it provides a way of critically theorising and investigating how new entities such as these are able to be created and *accounted for*, such that they can assume 'market values'. As stated above, our research draws closely on the performative turn in economics and economic sociology. This means that rather than exploring the existence and implications of an exogenous 'economic x' (as in conventional economics), we focus on *how* 'economic x is made'. In our case studies, we are applying the idea of a marketisation process, to understand the way that a value comes into being, and why it may, or may not, end in pricing, and possibly further financialisation. Part of what we do collaboratively is to show how newly created, quantified and variously marketised value entities have emerged, so that we can draw comparisons and conclusions across our different case studies about how values are made *in situ*. An initial typology of the range of calculative technologies in use can be made by broadly distinguishing them by what they produce. These outcomes include quantified *numeric values* that may become priced and financialised (e.g. carbon offset credits), *notional values* as computed in statistical and management tools (e.g. development impact assessments or quality of life adjusted years), and more qualitative *value framings* that frame and create the social and/or ecological categories and relationships that facilitate their formal consideration in numeric or notional terms.

In these categories of calculation technology a calculative device is usually deployed so as to settle and assign value. At the same time, and as highlighted in the notion of 'value framing', the process of value assignation is also influenced by social stigmas and cultural perspectives. It may include various forms of prejudice such as racism, misogyny and homophobia, as well as (more positively) being guided by principled and codified frameworks shaped by discourses regarding rights, humanitarianism and environmental care. All these influences play a part in the emergence of identifiable 'value entities' in our cases. These are the social framings or social categories that can be assigned to a person or an environmental unit, such that they become treated as, for example, an 'aid recipient', 'project affected person', 'orphaned and vulnerable child', 'HIV/AIDs affected person', 'biodiversity offset unit', 'carbon credit' and so on. To clarify, then, it is not the person or unit in themselves who becomes the entity that is valued, but the social or environmental category that has been formed through the processes described above, which may then attract an income stream, such as income from a development intervention, or be assigned a financial cost to be minimised.

It becomes quickly clear, however, that temporal, spatial and virtual processes can rapidly change the type of calculative technology in use. Moreover, movement between and among these various different calculative technologies can produce, in turn, a valued entity with quite different characteristics. These movements between categories are conditioned by actors who may encourage, resist or attempt to re-frame the valuation process. Our chapters illustrate as well these fluid, moving frontiers and the contested and unclear value renderings to which they may give rise.

In each of our case studies, then, we have researched the social articulation of value by focusing on the importance of emergent value entities in terms of:

- *how* these new value entities come into being in each of our case studies;
- *what* (calculative and value) assumptions are required on the part of their protagonists to enable the creation of specific value entities. For example, the possibility of creating a marketable biodiversity offset for the mitigation of environmental harm requires a host of assumptions and constructs regarding 'the nature of nature', accompanied by a whole other set of assumptions regarding things like development pathways, what it means to be human, the appropriateness of markets for best allocating environmental health and harm, etc. Similarly, in the case of water valuation, there appears to be an overarching need to create the notion of scarcity in order to introduce pricing, even in contexts where material scarcity is not apparent;
- the *implications* of these new circulating and value-accumulating entities for the problems they are constructed to address/redress, for the distribution of wealth and for the design of progressive policy.

We thus arrive at a core research question that concerns how calculative rationality conditions the types of calculative devices and technologies in use, and how the social articulation of these, in the full context of place, leads to emergent value entities with differing characteristics. We explore these processes as moving subjects, where the nature of the relationships between notional value and its framing, social categorisation, commodification and financialisation and how these processes work together and in contradiction creates contested and morphing value entities. Since calculative devices and technologies do not exist as *deus ex machina*, but arise in social processes, a key insight from the literature on socio-technical arrangements (which are specific, material and empirically researchable configurations of people, devices and things) is the need to understand the institutional arrangements in which calculations are made. The first means to do this is to map these institutional arrangements, and thus the second key node of our original research protocol was to recognise the importance of institutional assemblages.

2. *Institutional assemblages and assembling*

In the study of value and valuation processes, the case for thinking about valuation in terms of the distributed agency of an assemblage has been well established, perhaps most prominently by Michel Callon (e.g. Callon 1998, 2007; Çalışkan and Callon 2010) in his call for attending to the socio-technical agencements (STAs) through which markets are formed. The word 'agencement', generally translated as 'assemblage' in anglophone social theory, indicates a coming together of things "which are simultaneously human and nonhuman, social and technical, textual and material—*from which action springs*" (MacKenzie et al. 2007, 14–15, emphasis added). This last clause is critical for the current discussion as it underlines the point that this coming together of diverse things is not simply one mode through which agency is possible, but that it is "a necessary and prior condition for any action to occur" (Braun 2008, 671; see also Dewsbury 2011). Thus in the concept of agencement the individual, purposeful human actor is not only displaced as the sole possessor of action and agency, but no action is considered possible without the dynamic articulation or assembling of a diverse arrangement of people, beyond-human natures, things, texts, technologies and so on. The agency of people and things, in other words, is necessarily relational.

Although some versions of assemblage thinking focus mainly on the tendency of assemblages towards emergence, multiplicity and indeterminacy, this is only part of the story (Anderson and McFarlane 2011, 124). Assemblages may also tend towards stabilisation, normalisation and repetition (Dewsbury 2011). In their foundational conceptualisation of the dynamics of assemblage, for example, Deleuze and Guattari (1987[1980]) identify a twinned movement of lines of becoming: on the one hand there are lines of articulation, involving the coming together of heterogeneous things in an encounter through which (material, metaphorical, virtual, etc.) territorialisation occurs; on the other, there are lines of flight through which things are dispersed and the territories established through encounters dissolved, or deterritorialised (Mennicken and Miller 2012). Not all assemblages exhibit equal tendencies towards articulation/territorialisation and flight/deterritorialisation. The particular way in which different assemblages come to be configured, moreover, may set up or constrain different capacities and possibilities for future action (e.g. Callon 2007, 320). Indeed, Foucault's concept of apparatus (*dispositif*) – which as Legg (2011, 131) suggests may be thought of as a type of assemblage "prone to ... re-territorialisation, striation, scaling and governing" – has been central to our understanding of the relevant assemblages in our various case studies. According to Foucault an apparatus comprises the:

> discourses, institutions, architectural forms, regulatory decisions, laws, administrative measures, scientific statements, philosophical, moral and philanthropic propositions ... assembled to address an 'urgent need' and invested with strategic purpose.
>
> (Foucault 1980, 194)

Through such stabilised assemblages/apparatuses the particular, emplaced relations of value can become stabilised, repetitive *and therefore appear* to be the inevitable result of external social or economic "structures" (Law 1994). What is key here, however, is that despite these appearances, as Hinchliffe et al. (2007, 260–261) remind us, "such stabilities are themselves achievements which require all manner of human and nonhuman actions" and, as such, are orderings that are not imposed by universal laws or abstract forces external to the relations through which they are enacted. Instead, orderings arise from the relational arrangements and practices both contained within and excluded from particular assemblages.

In thinking about the different possibilities of assemblages as prone to deterritorialisation, immanence and lines of flight as well as to 're-territorialisation, striation, scaling and governing', it is useful to consider Manuel De Landa's (2009) figuring of assemblages as 'possibility spaces': dynamic relational spaces ordered by different topological rules that set the conditions for what forms and actions are possible. These conditions are not determined in advance, nor are they inevitable or immovable once in place. In contrast to essentialised notions of space with set properties:

> the capacities of an assemblage are not given, that is, they are merely possible when not exercised. But the set of possible capacities of an assemblage is not amorphous, however open-ended it may be, since different assemblages exhibit different sets of capacities.
>
> (De Landa 2009, 29)

Through this line of reasoning, the assemblages that constitute new markets and market logics for things previously outside of pecuniary systems of valuation can be said to set the possibility spaces of valuation in terms of economic capacities for calculating value as price. Indeed, arguably neoliberal governmentality can itself be seen as an assemblage through which:

> [t]he surface of contact between the individual and the power exercised on him, and so the principle of the regulation of power over the individual, will be only this kind of grid of homo economicus. Homo eonomicus is the interface of government and the individual.
>
> (Foucault 2008, 252–253)

3. Discursive framing

The third key dimension of our research protocol emphasises the *discursive framing providing* the systematic metaphorical "authorising knowledge" (Li 2007a, 2007b) for the acts and processes or calculation and assembling highlighted above. Every socio-technical arrangement (in Callon's terms) made up of the calculative technologies and devices shaped and applied through institutional arrangements also has an important, though frequently intangible, component guiding the ways that actors understand what is going on, their activities and behavior, and how they frame (Lakoff 2010) and understand valuation processes as a whole.

This link between knowledge and power and the idea that understanding is malleable by political processes has long informed political economy and manifested itself in Gramscian notions of hegemony. Over the past half-century it has become widely applied to investigation of the role of science in society via science and technology studies (Jasanoff 2004) but particularly in the contested arenas of environmental science and the analysis of how competing scientific opinions are constructed by discourses promoted through knowledge alliances (Hajer 1995, 2005; Stott and Sullivan 2000; Dryzek 2013). Watts and Peluso (2013) identify environmental discourses as a key part of 'regimes of truth' through which particular visions of environment and society are both formulated and legitimated. Thus, how one defines, or frames, social and environmental problems and goals will set the bounds of what are appropriate pathways to 'sustainability' (Scoones 2016) and through which measures (calculative technologies) we should assess progress along the chosen path.

Tensions and tendencies in performing value

The ways in which assemblages order socio-ecological spaces – and their differing capacities for doing so – can be expressed through the concept of performativity. Although economisation has been a central topic for exploring performativity, the concept is more widely relevant to studies of valuation of various types (see Çalışkan and Callon 2009, 2010). As Muniesa explains, central to performativity is the "idea of signification as act (to signify is an active process) and of reality as effectuation (to effect is to bring reality about)", two ideas he links to the pragmatist tradition in philosophy (Muniesa 2014, 16). Central to pragmatism and to performativity thus formulated is a close attention to situated practices. Muniesa points in particular to John Dewey's studies of value (1915a, 1915b, 1923, 1939), wherein he argued that value cannot be understood as a thing unto itself but instead can only be observed as a quality assigned to things (see above). That is, if something has a value (regardless of the register of that value – e.g. pecuniary or non-pecuniary) it is only because it is actively being valued. In short, valuation (as signification) is an act and the action of valuating things (achieved through the distributed agency of assemblages) effects a reality of value.

For Callon (1998) performativity signals the ways in which economics and economists do not describe a pre-existing reality – 'The Economy' – but instead participate in its making, through their implication in the formatting of the relations between elements within assemblages that constitute markets and other economic entities (Callon 1998; MacKenzie 2009; MacKenzie et al. 2007; Mitchell 2008; Muniesa 2014). In other words, the field of economics does not simply observe and describe the economy, but rather "performs, shapes and formats the economy" (Callon 1998, 2; Fourcade 2011; see also MacKenzie 2006). Importantly, Callon's work in performative economics highlights the point that performativity is not a one-off achievement, but involves ongoing movement – ongoing ordering and reordering. As Butler (2010, 149) explains in her reflection on economic performativity, the apparent separation of the economy from society does not happen just once (it is not locatable in history), such that "to say that the market is performatively produced is not to say that it is produced *ex nihilo* at every instant", but rather the performance of this separation involves ongoing reiteration. Once again we are reminded that arrangements that appear stable "are themselves achievements which require all manner of human and non-human actions" in each moment (Hinchliffe et al. 2007, 260–261).

One area of work that has been central to thinking about the performativity of economics relates to economisation and its sub-process of 'marketisation'. Economisation refers to the general framing of certain things and processes as formally economic – a process central to all of our case studies. Marketisation, meanwhile, refers to a specific version of economisation through which markets – and all of the physical and ideational infrastructure they require – are performed. Critically, marketisation involves defining and valuing entities that can be traded (Çalışkan and Callon 2010), actions that are especially relevant to our cases of new markets for water, carbon and biodiversity. In the process of this defining and valuing, a divide is enacted and reproduced between "the 'things' to be valued and the 'agencies' capable of valuing them" (ibid. 5). As all entities are understood here to have certain capacities depending on their relational entanglement with other entities, those that are to be valued for exchange in this divide must first have their agencies 'pacified', such that they can be transferred as property (ibid.). On the other side of this enacted divide are those entities with the capacity for calculating the value of that which is pacified (ibid.). Which entities are pacified and which are enacted as capable of pacifying is a matter of the unequal distribution of power within market socio-technical assemblages (STAs), as explored in greater detail below. For now, however, we reinforce the point made above that assemblages, including the STAs through which markets are enacted, may be formatted in certain ways to constrain some of the capabilities they contain. Again, assemblages can be both lively, unpredictable sites of transformation and sites of stabilised, ordered relations.

In the case of markets, the stabilisation and reproduction of relations ordered according to certain economic logics is performed by entities that successfully assert their ability to calculate values and thereby to define other, pacified, entities. Thus when it comes to market encounters, Çalışkan and Callon (2010, 14) write that:

> markets involve a series of multiple encounters and overlapping processes of calculations. Contingencies certainly play a part, as do the initiatives taken by agencies and the unpredictable movements of goods which overflow and follow unexpected trajectories. Yet encounters are not produced haphazardly. Like goods and agencies, they are also framed and formatted by a series of devices.

Chief among the various devices at play in processes of marketisation are calculative ones. Calculative devices – mathematical equations, scoring systems, indicators and so on – play a central role in the creation of new markets and the application of market logics to 'non-market' fields of activity, as shown in the case studies that follow. Calculative devices, then, effectively pacify things by formatting them into potentially equivalent units that can be exchanged.

Just as there is a danger of overemphasising the dynamic, emergent tendencies of assemblages however, there is also a danger of overemphasising stability and repetition in market STAs. In order to achieve scalable economic values, many things – unique aspects of particular ecosystems, the social and political contexts of development projects, individual human subjectivities – must be excluded from the calculative and discursive frame. For example, the STA to produce a new road or mall often acts to exclude persons who may informally live on the affected land or use it for grazing, and while the attempt is made to exclude these people from value considerations, they can often disrupt the STA and the construction process if they choose to. While this is an obvious example, STAs order, limit and privilege different types of knowledge and practice in complex ways. While STAs enact and perform certain calculative agencies capable of producing scalable effects and formatting relations according to market logics, in order to do so, certain things need to be emphasised within the frame of markets and others left out of this frame (for more on this structuring significance of framing, see Sullivan and Hannis 2015). As Hinchliffe and colleagues (2007, 272) contend, STAs striate spaces and format the relations they enact such that "certain things gain in significance and other things can drop out of the frame of reference". In conventional economics, those entities which are left out of market calculations are referred to as externalities. The economic imagination of externalities is one that revolves around the categories of costs and benefits: if something is left out of a market it is either a cost or a benefit to some party outside that market. In recognition that many of the things left out of markets cannot be readily understood as either costs or benefits – indeed, by virtue of their being left out of the frame many have not yet been subject to the processes of economisation by which they would be framed as such – Callon deploys a new term: 'overflows' (e.g. 2007). Those things that drop out of the frame of reference, or are purposefully left out of the frame (as is often the case, cf. Igoe 2014), have not been incorporated and thereby pacified. They therefore maintain their prior capacities to affect others and produce effects in the world, as for example in the effects of waste disposal by humans, the impacts of which are mediated by ecological processes. As such, overflows can become sites of lively agencies that, on occasion, can destabilise the very markets they were framed out of to begin with, so as to have 'counterperformative' effects (MacKenzie 2006; Nel 2015). Historically, the poor and vulnerable, non-humans and non-protected natures have often been consigned to the 'overflow' category of development and market capitalist STAs, which is a reason why valuation processes offer manifest opportunites for resistance to unjust marketisation.

One source of this counterperformativity is that overflows are often non-scalable: they cannot be expanded without the transformation or adjusting of elements and/or the relations between them (Tsing 2012). Indeed, this is often why they are left out of the frame to begin with. Another, sometimes related, source of counterperformativity is that overflows may be emergent, their arrival not predicted or predictable in advance, for example those that arise from unstable social conditions (e.g. Nel 2015) or from the irreducible complexity of ecosystems (Braun 2008). A third involves the economisation of externalities such that they become reconstituted within markets, as has occurred with the application of market-based mechanisms for resolving allocation issues in the distribution of environmental harms.

Despite the potentialities of overflows eventually becoming destabilising, however, new markets and market logics at the frontiers of valuation appear to be gaining, not losing, momentum. The indefinite nature of species and ecological relationships is almost completely absent from high level policy debates on biodiversity offsetting, which focus instead on developing ever more complex technical equations for establishing the commensurability of habitats. Increasingly dramatic evidence of anthropogenic climate change in the form of dramatic weather-related catastrophes (e.g. storms, floods, droughts and so on) and consensus that not enough is being done to slow its progress, has yet to destabilise carbon markets (albeit that they are currently characterised by low prices, turnover and liquidity). Similarly, criticisms of the rise of market logics in international humanitarian aid have not led to calls to find a new logic but rather to calls for improved planning and measurement technologies that include beneficiary voices (Krause 2014). The ability of valuation assemblages to resist or absorb any counterperformative effects from the overflows they create indicates a certain robustness derived from the unequal power relations within and between various assemblages.

If the objects of valuation practices have such lively capacities for overflowing the frames of marketisation, then, why is it that the latter not only seem to persist, but are seemingly ever advancing? Why is it that the representation of value in economic or market terms "supports some stakeholders' interests, while obscuring and abjecting those of others" (Bracking et al. 2014, 4)? To answer these questions, it is necessary to attend to the matter of power. Although agency is distributed through heterogeneous assemblages, power is not distributed equally within or between assemblages (Bennett 2010). As Marion Fourcade has demonstrated, the mere existence of an economic theory, model or calculative device does not mean it will have performative effects (Fourcade 2011). Some calculative devices are more successful than others, often as a result of their enrolment in powerful, i.e. empowered, STAs. An STA with more advanced calculative capacities and/or strong institutional involvement will have more power to define value(s) than one with relatively poor calculative capacities or weak institutional involvement. Thus inequalities in markets result, as Çalışkan and Callon (2010, 13) put it, from "the unequal power of calculating agencies that loop back to reinforce themselves. Due to these asymmetries, the most powerful agencies are able to impose their valuations on others and consequently to impact strongly on the distribution of value".

Such asymmetries are evident where the logic of markets is pushed ever further into the frontiers of valuation, framing aspects of human life and the non-human world not previously subject to market values in terms of price and costs. Those aspects of the people, places and things to be assigned economic values that resist such representation (for example because they are nonscalable in nature) are framed out of calculations, becoming overflows. In some instances these overflows may threaten the stability of economic valuation systems, as when it becomes obvious that a particular species will suffer from offsetting or the impacts of a development project will include clear harms and, as a consequence, sufficient resistance is generated to challenge the valuation systems producing these effects. In such instances overflows can become counterperformative. But in such cases the STAs producing economic values also tend to respond iteratively. The economic values they had previously produced do not simply dissolve in the face of resistance, but instead are adapted, or else new, more effective calculative technologies developed to (re)legitimise these values and thereby manage overflows more effectively and evade their potentially counterperformative effects (Callon 2007).

It is important to note here that the drivers of an STA's iterative responses to counterperformativities are not always internal to the existing STA, but may also include new aspects, including collective affects or 'affective atmospheres' (Anderson 2012). For example, collective anxieties about climate change (Weintrobe 2013) and feelings of loss related to extinction (Yusoff 2012) have contributed to demands that environmental degradation – whether through greenhouse gas emissions or more immediate forms of environmental destruction – be accounted for in economic terms. These collective affects become entangled with neoliberal discourses which hold that if things are not assigned economic values then there is no way to prevent their degradation. In other words, if things do not have economic value, they run the risk of being worthless in economic terms, even though they may be priceless in terms of intrinsic and aesthetic values (see Sullivan 2014).

Affective life and the vital, material and historically specific capacities of non-human materials – as both target and condition for the (bio)politics of market valuation – thus become entangled in neoliberal market STAs that frame them in terms of cost and price. This is so whether they are enrolled in these STAs as pacified commodities for exchange in markets, as indicators to be circulated amongst policy-makers, some mix of the two, or even if they are excluded from the STA as a market 'externality' – a cost for some other party outside the market being constructed to bear. Power plays an important role in these framings, giving some entities the ability to reinforce preferred striations to the exclusion of others or of lines of flight that escape striation. This is not as simple as the power of some vague and universal external force. Rather, power is likely to take subtly different, specific forms co-produced through the distributed agency of diverse assemblages that act in different ways in different instances of valuation. The case research shared in the chapters that follow documents this complexity for a range of valuation circumstances in development, conservation and environmental policy domains.

Conclusion

In this chapter we have briefly traced the historical and contingent nature of the contemporary dominance of economic forms of valuation in performing value. We have reviewed the diverse ways in which traditional approaches to valuation are found inadequate from a conceptual and practical standpoint . In response, and drawing strongly from economic sociology, political economy, anthropology, science and technology studies and ideas of performativity, we have identified an approach that draws on recent writing on the interplay between knowledge and power and the ways these establish key parameters of valuation systems: calculative technologies, institutional assemblages and framing discourses. In the following chapters we set out the results of a series of empirical studies of valuation practices in environmental and development contexts before attempting to draw together the lessons for economic valuation and its alternatives.

Note

1 This definition pertains to value the noun. The verb 'to value' is defined by the OED as "To estimate the value of" (see www.oed.com/search?searchType=dictionary&q=value&_searchBtn=Search).

References

Anderson B. (2012) "Affect and biopower: towards a politics of life" *Transactions of the Institute of British Geographers* 37: 28–43.

Anderson B. and McFarlane C. (2011) "Assemblage and geography" *Area* 43: 124–127.

Bakker K. and Bridge G. (2006) "Material worlds? Resource geographies and the 'matter of nature'" *Progress in Human Geography* 31: 5–27.

Bennett J. (2010) *Vibrant matter: a political ecology of things* Duke University Press, Durham, NC.

Bichler S. and Nitzan J. (2015) "Capital accumulation: fiction and reality" *Real-World Economics Review* 72: 47–68.

Boltanski L. and Thévenot L. (2006[1991]) *On justification: economies of worth* (trans. C Porter) Princeton University Press, Princeton, NJ.

Bourdieu P. (1993) *The field of cultural production* Columbia University Press, New York.

Bracking S., Bond P., Brockington D., Büscher B., Igoe J., Sullivan S. and Woodhouse P. (2014) *Initial research design: 'human, nonhuman and environmental value systems: an impossible frontier?'* LCSV Working Article Series, no. 1. University of Manchester, Manchester.

Bracking S. and Ganho A.S. (2011) *Investing in private sector development: what are the returns? A review of development impact evaluation systems used by development finance institutions in Europe* Norwegian Church Aid, Oslo.

Braun B. (2008) "Environmental issues: inventive life" *Progress in Human Geography* 32: 667–679.

Butler J. (2010) "Performative agency" *Journal of Cultural Economy* 3(2): 147–161.

Çalışkan K. and Callon M. (2009) "Economization Part 1: shifting attention from the economy towards processes of economization" *Economy and Society* 38: 369–398.

Çalışkan K. and Callon M. (2010) "Economization Part 2: a research programme for the study of markets" *Economy and Society* 39: 1–32.

Callon M. (1998) "Introduction: the embeddedness of economic markets in economics" in Callon M. ed., *The Laws of the Markets* Blackwell Publishers/Sociological Review, Oxford, 1–57.

Callon M. (2007) "What does it mean to say that economics is performative?" in MacKenzie D., Muniesa F. and Siu L. eds, *Do economists make markets? On the performativity of economics* Princeton University Press, Princeton, NJ.

Castree N. (2002) "False antithesis? Marxism, nature and actor-networks" *Antipode* 34: 111–146.

Castree N. (2003) "Environmental issues: relational ontologies and hybrid politics" *Progress in Human Geography* 27: 203–211.

Coase R. (1960) "The problem of social cost" *Journal of Law & Economics* 3: 1–44.

Corson C., MacDonald K.I., Neimark B. (2013) "Grabbing green: markets, environmental governance and the materialization of natural capital" *Human Geography* 6(1): 1–15.

Cronon W. (1991) *Nature's metropolis: Chicago and the Great West* Norton, New York.

De Landa M. (2009) *A new philosophy of society: assemblage theory and social complexity* Continuum, London.

Deleuze G. and Guattari F. (1987[1980]) *A thousand plateaus: capitalism and schizophrenia* (trans. B. Massumi) Continuum, London.

Dewey J. (1915a) "The logic of judgments of practise (Part I)" *Journal of Philosophy, Psychology and Scientific Methods* 12(19): 505–523.

Dewey J. (1915b) "The logic of judgments of practise (Part II)" *Journal of Philosophy, Psychology and Scientific Methods* 12(20): 533–543.

Dewey J. (1923) "Values, liking and thought" *Journal of Philosophy* 20(23): 617–622.

Dewey J. (1939) *Theory of valuation* University of Chicago Press, Chicago, IL.

Dewsbury J-D. (2011) "The Deleuze–Guattarian assemblage: plastic habits" *Area* 43: 148–153.

Dryzek J. (2013) *The politics of the earth: environmental discourses* Oxford University Press, Oxford.

Dussauge I., Helgesson C-F. and Lee F. (eds) (2015) *Value practices in the life sciences and medicine* Oxford University Press, Oxford.

Federici S. (2004) *Caliban and the witch* Autonomedia, New York.

Foucault M. (1980) *Power/knowledge: selected interviews and other writings 1972–1977* (C. Gordon ed.; trans. C. Gordon, L. Marshall, J. Mepham and K. Soper) Harvester Press, Brighton.

Foucault M. (1991[1975]) *Discipline and punish: the birth of the prison* (trans. A Sheridan) Penguin, London.

Foucault M. (2007) *Security, territory, population: lectures at the Collège de France, 1977–1978* Palgrave Macmillan, New York.

Foucault M. (2008) *The birth of biopolitics: lectures at the Collège de France, 1978–1979* Palgrave Macmillan, New York.

Fourcade M. (2011) "Cents and sensibility: economic valuation and the nature of 'nature'" *American Journal of Sociology* 116: 1721–1777.

Gareau B.J. (2005) "We have never been human: agential nature, ANT, and Marxist political ecology" *Capitalism, Nature, Socialism* 16: 127–140.

Gorman H.S. and Soloman B.D. (2002) "The origins and practice of emissions trading" *Journal of Policy History* 14: 293–320.

Graeber D. (2001) *Toward an anthropological theory of value: the false coin of our own dreams* Palgrave MacMillan, Basingstoke.

Hajer M.A. (1995) *The politics of environmental discourse: ecological modernization and the policy process* Oxford University Press, Oxford.

Hajer M.A. (2005) "Coalitions, practices, and meaning in environmental politics: from acid rain to BSE" in Howarth D. and Torfing J. eds, *Discourse theory in European politics* Palgrave Macmillan, London.

Henderson G. (2013) *Value in Marx: the persistence of value in a more-than-capitalist world* University of Minnesota Press, Minneapolis.

Hinchliffe S., Kearnes M.B., Degen M. and Whatmore S. (2007) "Ecologies and economies of action: sustainability, calculations, and other things" *Environment and Planning A* 39: 260–282.

Holloway J. (2015) "Read capital: the first sentence. Or, capital starts with wealth, not the commodity" *Historical Materialism* 23: 3–26.

Igoe J. (2014) "Firewall" in Fredriksen A. et al. *A conceptual map for the study of value* LCSV Working Article Series, no. 2. University of Manchester, Manchester, 35.

Jasanoff S. (2004) *States of knowledge: the co-production of science and the social order* Routledge, London.
Krause M. (2014) *The good project: humanitarian NGOs and the fragmentation of reason* University of Chicago Press, Chicago, IL.
Lakoff G. (2010) "Why it matters how we frame the environment" *Environmental Communication* 4: 70–81.
Lamont M. (2012) "Toward a comparative sociology of valuation and evaluation" *Annual Review of Sociology* 38: 201–221.
Lane R. (2012) "The promiscuous history of market efficiency: the development of early emissions trading systems" *Environmental Politics* 21: 583–603.
Law J. (1994) *Organizing modernity* Blackwell, Oxford.
Legg S. (2011) "Assemblage/apparatus: using Deleuze and Foucault" *Area* 43: 128–133.
Li T.M. (2007a) "Practices of assemblage and community forest managements" *Economy and Society* 36(2): 263–293.
Li T.M. (2007b) *The will to improve: governmentality, development, and the practice of politics* Duke University Press, Durham, NC.
Luxemburg R. (2003 [1913]) *The accumulation of capital* Routledge, New York.
MacKenzie D. (2006) *An engine, not a camera: how financial models shape markets* MIT Press, Cambridge, MA.
MacKenzie D. (2009) "Making things the same: gases, emission rights and the politics of carbon markets" *Accounting, Organizations and Society* 34: 440–455.
MacKenzie D., Muniesa F. and Siu L. (2007) *Do economists make markets? On the performativity of economics* Princeton University Press, Princeton, NJ.
Mandelbrot B. (2008) *The (mis)behaviour of markets: a fractal view of risk, ruin and reward* Profile Books, London.
Marx (1993[1857–1858]) *Grundrisse: foundations of the critique of political economy* Penguin Classics, London.
Meadows D.H., Meadows D.L., Randers J. and Behrens W.W. III (1972) *The limits to growth* Potomac Associates, Washington, DC.
Mennicken A. and Miller P. (2012) "Accounting, territorialization and power" *Foucault Studies* 13: 4–24.
Mitchell T. (2008) "Rethinking economy" *Geoforum* 39(3): 1116–1121.
Muniesa F. (2012) "A flank movement in the understanding of valuation" *Sociological Review* 59(s2): 24–38.
Muniesa F. (2014) *The provoked economy: economic reality and the performative turn* Routledge, London.
Nel A. (2015) "The neoliberalisation of forestry governance, market environmentalism and re-territorialisation in Uganda" *Third World Quarterly* 36(12): 2294–2315.
Nelson S. (2014) "Resilience and the neoliberal counterrevolution: from ecologies of control to production of the common" *Resilience* 2(1): 1–17.
Nelson S. (2015) "Beyond the limits to growth: ecology and the neoliberal counter-revolution" *Antipode* 47(2): 461–480.
O'Neill J. (2007) *Markets, deliberation and the environment* Routledge, London.
Pearce D., Markandya A. and Barbier E. (1989) *Blueprint for a green economy* Earthscan, London.
Peters C.M. (1996) *The ecology and management of non-timber forest resources* World Bank, Washington, DC.

Robertson M. (2004) "The neoliberalization of ecosystem services: wetland mitigation banking and problems in environmental governance" *Geoforum* 35: 361–373.
Scoones I. (2016) "The politics of sustainability and development" *Annual Review of Environmental Resources* 41: 293–319.
Skeggs B. (2014) "Values beyond value? Is anything beyond the logic of capital?" *British Journal of Sociology* 65: 1–20.
Smith N. (1984) *Uneven development: nature, capital and the nature of space* Blackwell, Oxford.
Stark D. (2009) *The sense of dissonance: accounts of worth in economic life* Princeton University Press, Princeton, NJ.
Stott P. and Sullivan S. (eds) (2000) *Political ecology: science, myth and power* Edward Arnold, London.
Strathern M. (2000) *Audit cultures: anthropological studies in accountability, ethics and the academy* Routledge, London.
Sullivan S. (2014) *The natural capital myth; or will accounting save the world? Preliminary thoughts on nature, finance and values* LCSV Working Article Series, no. 3. University of Manchester, Manchester.
Sullivan S. (forthcoming) "Making nature investable: from legibility to leverageability in fabricating 'nature' as 'natural capital'" *Science and Technology Studies* Online First at https://sciencetechnologystudies.journal.fi/forthcoming/article/58040/27301
Sullivan S. and Hannis M. (2015) "Nets and frames, losses and gains: value struggles in engagements with biodiversity offsetting policy in England" *Ecosystem Services* 15: 162–173.
Swyngedouw E. (1996) "The city as hybrid: on nature, society and cyborg urbanization" *Capitalism Nature Socialism* 7: 65–80.
Swyngedouw E. (1999) "Modernity and hybridity: nature *Regeneracionism* and the production of the Spanish waterscape 1890–1930" *Annals of the Association of American Geographers* 89: 443–465.
Taussig M. (2010[1980]) *The devil and commodity fetishism in South America* University of North Carolina Press, Chapel Hill.
Tsing A.L. (2012) "On nonscalability: the living world is not amenable to precision-nested scales" *Common Knowledge* 18(3): 505–524.
Watts M. and Peluso N. (2013) "Resource violence" in Death C. ed., *Critical environmental politics* Routledge, London, 194–197.
WCED [World Commission on Environment and Development] (1987) *Our common future* Oxford University Press, Oxford.
Weber M. (1930[1905]) *The Protestant ethic and the spirit of capitalism* (trans. T. Parsons) George Allen & Unwin, London.
Weintrobe S. (ed.) (2013). *Engaging with climate change: psychoanalytic and interdisciplinary perspectives* Routledge, London.
Yusoff K. (2012) "Aesthetics of loss: biodiversity, banal violence and biotic subjects" *Transactions of the Institute of British Geographers NS* 37: 578–592.

Part 1

Development

3 Assembling value for money in the UK Department for International Development

Aurora Fredriksen

Introduction

In 2010 the UK's newly elected coalition government began enacting a wide-ranging programme of austerity, involving drastic budget cuts across all government departments except the National Health Service (NHS) and the Department for International Development (DfID). While sparing the NHS from financial cuts was widely popular, the ring-fencing of DfID's budget was less so. This decision drew negative attention amid press allegations that DfID was wasting taxpayers' money on excessive overhead costs (e.g. Boin 2010; Dixon 2010) and surveys showing a decline in public support for foreign aid (TNS UK 2010). In this context, the contemporaneous mandate that all government departments demonstrate 'value for money' for all spending gained particular traction at DfID, where it became, according to one policy consultant "a sort of mantra"[1] for defending continued spending on international aid at a time of domestic austerity. At the same time that demonstrating 'value for money' became a means of justifying the worth of spending taxpayers' money on international development, it also drew criticism from those who worried that its requirement to show measurable results would discourage certain types of aid and replace (or obscure) difficult political and moral decisions with technocratic procedure (e.g. Leroy 2011; Scott et al. 2012). Taking this concern seriously, in this chapter I look at *how* value for money is calculated and framed in practice, and how these practices *perform* the value of international development programming in particular ways, to the exclusion of alternatives.

It should be noted that, while this chapter takes UK DfID as a site for exploring the demonstration of value for money (hereafter VfM) in development aid, the use of VfM calculations in development assessments and evaluations is by no means unique to DfID or the UK. Indeed, demonstrating the VfM of development aid is now common to several large donors (Bond 2011; Davis 2012). Consequently, while this chapter draws its empirical material from interviews and documentary reviews centred on VfM at UK DfID, it also speaks to wider issues around VfM as a measure of worth in international development aid more generally. The chapter begins by briefly situating VfM within the larger institutional setting of development aid in the post-Second World War era and, in particular, the rise of the aid effectiveness agenda from the 1990s onwards. This is followed by an examination of both the calculation and the framing of VfM at DfID. An analysis of the performative effects of these practices supports the conclusion that VfM does not simply capture an objective measure of development value, but instead performatively shapes what counts as development value in the first place. A concluding section returns to the concern introduced above, reflecting on the potential for technocratic justifications like VfM to displace moral and political engagements, debate and justifications for development decisions.

Situating value for money

As with other fields of policy and practice (see Chapters 1 and 2, this volume), in international development the generation, assessment and enactment of various registers of value – e.g. economic, political, moral – have never been easily disentangled from one another (though neither have they been coterminous). At least since its modern inception following the Second World War, international development aid has emphasised economic value, largely understood in the 20th century in terms of macro-economic growth as measured by gross domestic product (GDP). At the same time, development aid has consistently been concerned with generating non-economic value(s), most notably geo-political value(s), but also moral or charitable value(s) and, more recently, social and rights-based value(s).[2] While arriving relatively recently on the international development scene, VfM is part of a broader shift in the locus of value assessment in international aid, commencing with the post-cold war rise of the related agendas of 'aid effectiveness', 'aid accountability', and 'new public management'.

The rise of aid effectiveness and accountability doctrines from the 1990s onward has had a major impact on how the value of international development aid is understood, measured, and projected. In response to the widespread failure of international development aid to lift countries out of poverty and generate sustained economic growth in the post-war decades, aid effectiveness – codified in the 2005 Paris Declaration on Aid Effectiveness – seeks to "improve the quality of aid and its impact on development" (OECD 2017), by deploying better assessment, monitoring, and evaluation techniques to make it (literally) more accountable. Thus the aid effectiveness agenda has had the practical consequence of replacing the broad and largely non-specific macro-goals of earlier regimes of international development, with a regime of specific accountability for results at the level of individual programmes and projects (Krause 2014). One outcome is that aid effectiveness, as a project to make aid more accountable, has in practice made different aid projects and programmes more comparable through practices that frame projects in certain standardised ways. Aid effectiveness standardises the requirement for projects and programmes across the board to be explicit about their goals, how these will be achieved, and how evidence of the effectiveness of the proposed project or programme will be demonstrated.

As well as this narrowing of the locus of development value (from the macro-economy to the project), the requirement that aid effectiveness show measurable results has had at least two notable unintended consequences. First, this makes broadly ambitious, innovative, or idealistic programmes and projects more difficult to implement (e.g. Mosse 2004, 2011). Second (reinforcing the first), it has brought about what Krause (2014) describes as a process of 'quasi-marketization' (cf. Çalışkan and Callon 2010) in which consultancies and non-governmental organisations (NGOs) compete for contracts from donors in a market where the beneficiaries of aid stand in as quasi-commodities. With its emphasis on cost savings, competition, and establishing a framework for comparing different projects and programmes, VfM extends and deepens the quasi-marketisation of development aid and its beneficiaries. I expand on this point below through examination of the calculation, framing, and performativity of VfM. First, however, it is useful to define more specifically what DfID means by the phrase 'value for money'.

Defining value for money

The UK National Accounting Office defines VfM as "the optimal use of resources to achieve intended outcomes". Building from this official definition, DfID has declared that, for them, VfM means "maximiz[ing] the impact of each pound spent to improve poor people's lives" (DfID 2011, 3). Moreover, DfID's 'Common Agenda for Development Results' (DfID 2011) calls for a re-orientation of its development policies (and those of its partners) that not only highlights results and VfM, but puts these 'at the heart' of everything it does. The treatment of VfM in these statements speaks to the 'mantra'-like status that VfM holds for DfID noted above. VfM became a way of defending the sparing of DfID's budget from austerity-driven funding cuts, for "saying ok, well we're committed internationally to raising the aid budget to 0.7% of GDP, but in doing this we will make sure that … every pound spent on aid will get the maximum amount of impact" (Interview 04/07/15). DfID is quick to note, however, that attaining good VfM is not about going with the cheapest options, but going for the most efficient ones, i.e. the cheapest option that still results in the desired outcomes. Thus VfM also speaks to a renewed emphasis on 'evidence-based' decision-making in international development. This is indicated for example, in statements from the Common Agenda report such as these:

> To maximise the impact of UK aid we need to be very clear about the results (outputs and outcomes) we expect to achieve as well as the costs.
> (DfID 2011, 3)

> For VfM agenda to succeed we all need to show diligence in what we decide to do and how we implement (knowing what works and being sure that we are delivering development results in the most effective and efficient way).
> (DfID 2011, 12)

These accounts, and moreover the idea of evidence-based decision-making in general, assume that it is possible to widely evidence 'what works' in the first place. Leaving aside this questionable assumption, the remainder of this chapter traces the processes of *doing* VfM, examining how VfM is calculated, framed, and performed.

Calculating value for money

There is no standard or prescriptive way of calculating VfM. Indeed, VfM is not itself a calculative device; it is not a formula or accounting tool, but rather a way of framing and performing value as a relationship between inputs and outputs or outcomes (e.g. Jackson 2012). Guidance notes produced by consultants contracted by DfID have varied by author and sector in the methods they use or recommend for calculating VfM. The most commonly used techniques for assessing VfM within aid agencies include: cost–benefit analysis (CBA), cost-effectiveness analysis (CEA), social-return on investment (SROI), and basic efficiency resource analysis (BER) (for an overview of each of these methods see Davis 2012). However, assessment techniques that are less familiar to the aid community such as Randomised Control Trials (RCTs) and community scoring have also been suggested (ICAI 2011; various interviews).

The method used to calculate VfM is thus flexible, more a matter of reaching an agreement between the donor and the consultant or delivery partner (though often to the notable exclusion of beneficiaries). Take, for example, descriptions of the process of deciding how to measure VfM from two different development consultants:

> You submit your proposal, there's an evaluation, and quite often there's a preferred bidder at that point ... then negotiations start in terms of, you've submitted this as your value for money approach, now we need to firm up on that, can you firm up on the criteria? How are you going to deliver value for money and enhanced value for money, you know, what's above and beyond what you've stated there that could be used in scale-up. So, it's an ongoing discussion, but it's a bit one-way in terms of they're expecting – quite rightly to a certain extent – the consultant to come up with ways on the programme that real value for money could be enhanced and added going forward.
>
> (Interview 22/10/15)

> The DfID staff member was concerned that we had sort of cut and dried guidance on what to measure, or metrics as he called them, everything had to be metrics for this and that.
>
> (Interview 04/07/15)

As these two accounts show, consultants are hired by large donors like DfID to come up with ways of assessing or demonstrating the VfM of different programmes or projects, and while donors will accept a wide variety of techniques, they want these to be clearly laid out and robustly justified in the consultant's report. In practice, this often comes down to a desire for quantitative measures based on an assumption (often, but not always, unspoken) that quantification equals robustness.

The institutional push to quantify outcomes – to have metrics for every development result – has unsurprisingly led to concerns that less quantifiable outputs or impacts might be side-lined in favour of more easily counted ones (e.g. Vardakoulias 2013). Consider, for example, the following scenario recounted by one interviewee:

> Let's say you're building six more classrooms and the result of those six more classrooms is six times 30 kids attend so you've got 180 extra kids every year attending that school; that's straightforward, you know that's very easy. But when you're talking about how the welfare and wellbeing and the economic situation in a district has improved as a result of access being opened up [from building a new road], that's much more difficult. That's part of the issue in terms of measurement: it's not easy to measure some things.
>
> (Interview 22/10/15)

Certain results (e.g. school attendance), can be easily quantified in VfM calculations, while others (e.g. wellbeing), cannot.

DfID's official line is that the difficulty of quantifying some types of outcomes or impacts should not be a hindrance to including them in VfM assessments. As stated in the Common Agenda: "It's very hard to quantify some benefits e.g. what is the value of a free and fair election? But demonstrating results and tackling difficult development issues are not alternatives – we need to do both" (DfID 2011, 6). In other words, consultants and service providers will need to work even harder to develop new ways of measuring and quantifying those more 'difficult-to-quantify' results. Notice then how quantifying benefits and demonstrating results are equated in the call to rise to the challenge of demonstrating (i.e. quantifying) the benefits of aid for addressing 'difficult development issues'.

This likening of quantification with 'evidence' is not unique to DfID, but widespread throughout international development and governance institutions, where reliance on statistics in general and indicators in particular is paramount (see Davis et al. 2010; Fredriksen 2012; Hansen and Porter 2012; Merry 2011). The appeal of quantification comes from the capacity of quantified data "to convert complicated contextually variable phenomena into unambiguous, clear and impersonal measures" (Merry 2011, S84), depicting them as transparent, non-interpretive – and therefore apolitical – facts (Porter 1995; Poovey 1998). By assuming the appearance of neutral fact, quantified representations of VfM allow policy-makers and desk officers at DfID to decide between diverse projects and programming in a way that seems fair, rational, and non-political, a feature that is common to other larger donors too.

In parallel, the process of producing quantified evidence of results is, in practice, always messier and more interpretive than its final numerical representation would suggest. Turning results into numbers to plug into CBA, CBE, SROI, or other equations for calculating VfM, inevitably involves subjective judgements, negotiations, and even 'gut feelings' on the part of the consultants, service providers, and desk officers involved. As with other types of calculation, the process of calculating VfM might be better described as one of 'qualculation' (Cochoy 2008), in which quantitative calculation is informed by qualitative judgements. This is reflected in the account of one consultant who described his involvement in developing VfM guidance in terms of an opening up of the concept to include qualitative judgements:

> It's not that I'm terrifically interested in doing endless appraisals or cost-benefit analyses. I'm quite sceptical of the whole idea actually. But for me it's been a route for being involved in [development] programmes and trying to make sure that the interpretation of value for money is sufficiently broad. See, one of the problems … is that it tends to be reduced to a few sort of standard formulae in DfID value for money work and not interpreted in a sufficiently broad fashion.
>
> (Interview 04/07/15)

Another consultant emphasised the role of expert judgement and 'gut feeling' in VfM calculations, explaining:

> As consultants we kind of work on evidence and gut feeling and relationships with stakeholders and you get a much better, sort of more rounded view. And some of the evidence [DfID are] looking for has to be for clear tangible evidence, whereas some of it is actually more relationships and understanding of stakeholder drivers, influences and those types of things.
>
> (Interview 22/10/15)

Undeniably, some official guidance on calculating VfM notes that translating qualitative realities into quantified measures will require assessors to make judgement calls; as one such report states, calculating VfM "will invariably involve large doses of art and subjective judgement" (ICAI 2011, 3).

The judgements and 'gut feelings' underlying the assessment of VfM (and indeed many other development assessments and evaluations) appear in official documents as stated assumptions. As with other fields where quantified accounting practices dominate, it is standard to make assumptions explicit when translating perceived development results into quantified development outcomes as a way of *accounting for judgement*. Thus using 'gut feeling and relationships' is acceptable in demonstrating VfM, so long as you "demonstrate how you got to that, which is then questionable of course" (Interview 22/10/15). In other words, the use of subjective judgements in VfM calculations is negotiated between desk officers at DfID and consultants or service providers. The following contractor's experience with project proposals speaks to this:

> I think quite often there's enough evidence to be able to make the move on [demonstrating the VfM of a project], but DfID may insist that, no it's not sufficiently evidence-based, it's not robust enough, it could be questioned, it could be whatever. And we do see that at times depending on different advisors.
>
> (Interview 22/10/15)

So, while VfM is "not the hard and rigorous process that the political rhetoric around it may suggest" (ICAI 2011, 3), there are limits to the use of personal judgements in quantifying VfM. DfID staff tend to be less comfortable when VfM assessments rely on 'too many' assumptions or when assumptions are based on personal experience and/or gut feelings alone. Assumptions are acceptable, but are best couched in 'evidence' from previous evaluations or accepted indicators. In other words, they should have some quantitative content, regardless of whether such content involved a high degree of expert judgement and negotiation (see Merry 2011). Again, we see 'quantitative' equated with 'robust' in the demonstration of VfM.

Calculating VfM, as this section has shown, is a process involving a variety of calculative techniques (CBA, CBE, SROI, BER, etc.), all of which involve at least some expert judgement and 'qualculations' in the translation of qualitatively messy situations into quantities of development outcomes. While they rely on a wide range of methods, all of these techniques have one thing in common: they aim to produce commensurable values for very different projects and programmes. The use of assumptions (preferably based on some pre-existing standards) allows the non-quantifiable particularities of situations to be framed out of VfM assessments, making them commensurable across different projects, places, and time frames. The cost is that such quantified results (and their relationship to demonstrating the VfM of a development project or programme), may exclude important aspects of a particular context, giving only a partial picture of an intervention's efficacy, or lack thereof. The process of quantification is, however, not the only way in which different projects and programmes are made commensurable through VfM assessments. In the following section, I look at the framing of VfM as a linear relationship expressible through the 'logframe' approach.

Framing value for money

Once the incommensurable qualitative realities of a project or programme have been translated into commensurable quantities of development outcomes as described above, they can be used to create a logframe (i.e. a logical framework matrix, a standard planning tool for showing how project inputs will translate into results, or outcomes). Among other things (see Krause 2014 on the history and use of logframes), the use of logframes figures a linear relationship between project inputs and development outcomes and impacts. In the case of VfM, projects are encouraged to use a modified logframe based on the '3Es' (see Figure 3.1). The 3Es logframe shows a clear linear movement from inputs to process to three successive levels of results: outputs, outcomes, and impact. Inputs can be easily costed, and thus account for the first 'E' – *economy*. The second 'E' accounts for how *efficiently* inputs are translated to outputs, while the third 'E' – *effectiveness* – accounts for how well outputs engender the desired outcomes. An arrow leading from inputs (economy) to impact represents 'cost-effectiveness', while a 'fourth E' – *equity* – is depicted as operating from the start to the finish of the chain. It is notable, however, that the framework continues to foreground the '3Es', thereby relegating this fourth E to a sort of afterthought.

The 3Es framework for framing VfM is based on DfID's more general logframe template, with its standard 'results chain' terminology (which replaces the terms 'goal' and 'purpose' common in other logframe schemas with 'impact' and 'outcome'). Of course, the very language of a 'results chain' implies linearity, conveying the sense of a chain moving from one

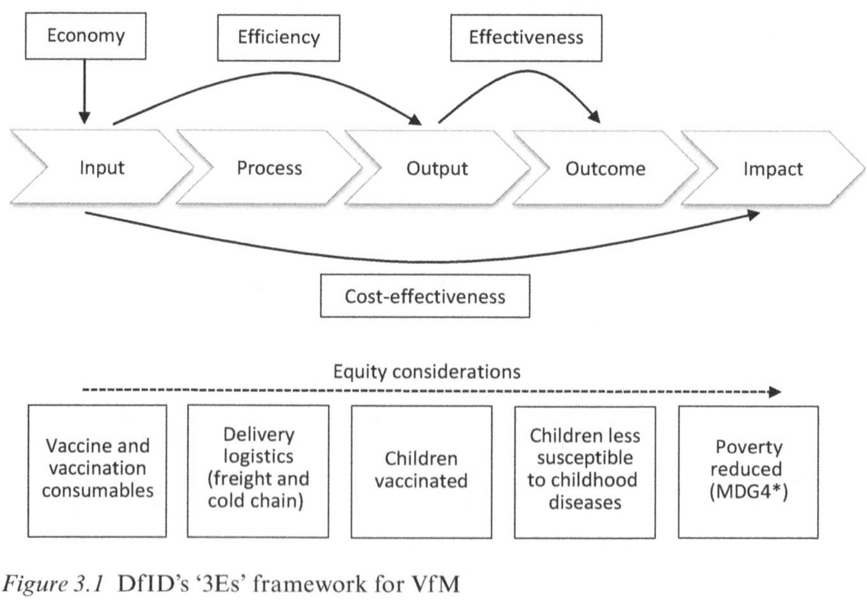

Figure 3.1 DfID's '3Es' framework for VfM

Source: DfID (2011, 4). In this case, DfID are illustrating the 3Es chain in relation to reducing child mortality.

Note

*MDG4 refers to the Millennium Development Goal 4: reduce child mortality.

link to the next. More broadly, the logframe model posits a clear progression from inputs to outputs and impacts in which project planners can clearly link inputs to results in a straightforward manner. Any evidence and assumptions used to support the claims made by the logframe must be clearly stated up front (see above) and adaptation of the model as the project unfolds is limited. Talking to this point, a development consultant for DfID explained:

> Whilst they [DfID] ask for value for money, they don't want to change too much. And part of that is linked to the milestone payments – payment by results – because they're all measured against the logframe which have clear and tangible outputs. And if you start to adapt the programme because you see opportunities elsewhere, you move away from that logframe, so it's difficult then for them to measure and demonstrate the performance of the programme, despite the fact it could actually have much greater impact as a result of that.
>
> (Interview 22/10/15)

Moving away from the logframe by adapting projects as they develop disrupts its linearity, hampering the ability to measure eventual impacts against initial inputs in the way the logframe requires.

In the same way that quantification can mask the details of specific contexts, the linearity of DfID's 3Es logframe for VfM obscures the more complicated, non-linear, relationships that often exist on the ground. Even when a desired impact (e.g. economic growth) is in itself relatively easy to measure, the extent to which the inputs of a specific project or programme contributed to this larger impact is often unknown. In the above example of opening up access, the inputs were roads. While there is reasonable evidence that opening up access to remote locations will contribute to economic growth, it is impossible to know how much this specific input – roads (or even its associated outcome – opened up access) – contributed to the larger impact of economic growth. This, after all, is likely to be affected by a wealth of non-linear, entangled relationships involving anything from local weather conditions to distant stock markets. Moreover, economic growth is unlikely to be evenly distributed across all groups in an area such that some groups of people could even see a decline in their economic standing that is hidden by aggregate measures of economic growth. When an impact (e.g. increased wellbeing) is difficult to quantify, the problem of assigning attribution is only amplified. The logframe, as Krause explains, "separates evidence of one specific project result from all other effects NGO work might be having" (2014, 89), not to mention the potential entanglement of other causes and effects in the larger context in which a project or programme takes place. Yet, by framing the relationship between inputs and outcomes/impacts in such a direct, linear fashion, the 3Es logframe *performs* VfM in a particular way, a process explored further in the next section.

The performativity of VfM

As noted above, VfM at DfID is both a means of guiding decision-making and a defence of decisions made by DfID against what one interviewee called "the *Daily Mail*s of the world". Thus VfM performs value first in the more literal sense of the word, as a demonstration of worth to justify spending decisions and defend against would-be critics. VfM can also be said to perform value in the more theoretical sense of being *performative* (as laid out in Chapters 1 and 2 this volume; see also, e.g. Callon 1998; MacKenzie 2003; MacKenzie et al. 2007; Mitchell 2002) i.e. as in fact effecting the reality it claims to describe (e.g. Muniesa 2014). VfM does not simply describe the relative value of different development projects and programmes, it shapes and enacts what counts as a valuable development project or programme in the first place. The remainder of this section expands on the notion of *performativity* as it relates to VfM.

Importantly, to say that something is performative – that it brings about the reality it claims to describe – does not necessarily entail a corresponding assumption that reality will straightforwardly and completely align with representations of itself (Fourcade 2011; Muniesa and Linhardt 2011). In the case of VfM, Krause's argument about logframes in general is apt: "Management tools like the logframe do not determine what people do, but they shape it: they shape what people get to see and know about the world, and people's ideas about what the task before them is" (2014, 76). VfM, with its requirements to quantify development outcomes and set them in a linear relationship to costed inputs using the 3Es logframe, shapes what desk officers, policy-makers, and others see and know about development projects – what gets counted and what gets excluded from the frame of reference. It consequently shapes ideas about what types of results matter (i.e. ones that can be quantified) and how projects and programmes should be conceptualised – as involving a linear relationship between inputs and results or outcomes. Critically, by setting these parameters, VfM performs itself as a ratio of costed inputs to quantified (and if possible monetised) outcomes. Value that falls outside of this formulation – that is not quantifiable or cannot be set into a neatly linear relationship – is left out of the performance.

Performativity, it should be noted, is not a one-off achievement but a matter of ongoing (re)iterations (Butler 2010, 149). Successive iterations of VfM assessment, within DfID as well as a wide range of donors, strengthen its ability to perform its constricted version of value, as more donors and contractors (e.g. NGOs) learn to shape their projects and programmes to deliver VfM as a ratio of costed inputs to outcomes. However in a broader sense, VfM as a set of practices can also be understood as a performative iteration of development value – as the quantified value of individual projects and programmes within the larger performativity of development economics and aid accountability. As one interviewee explained:

> One thing about all this VfM everyone talks about is that there's nothing new about it. It's what [development] economists have been doing for

> many decades: it's basically cost-effectiveness analysis, cost efficiency, cost-effectiveness, and cost-benefit analysis. It's those things that used to be standard in project appraisals... the difference is that it has to be more explicit.
>
> (Interview 04/07/15)

In this quote, the interviewee – a freelance development consultant who had worked with DfID, the World Bank, and other aid agencies for several decades – states that "there's nothing new" about value for money, likening it to previous modes of assessing the value of aid at the level of individual projects and programmes. Indeed it is true that VfM did not introduce economic accounting methods to international aid at the level of the individual project, a trend associated with the larger aid effectiveness, aid accountability, and the new public management agenda of the 1990s. At the same time, VfM reinforces earlier iterations of development value as a linear economic relationship between quantified inputs and outcomes; it makes the parameters of development value both narrower and 'more explicit', bringing them to the front and centre of decision-making. Making things explicit, however, as Muniesa and Lindhardt (2011) point out, can bring new questions to the fore and open up new, previously hidden, grey areas. This discussion on calculating VfM has showed how in the process of reinforcing the need to use certain types of (quantitative) evidence, VfM opens up spaces for negotiation, judgement, and even 'gut feelings'. These openings, however, should not be overstated. Negotiation and judgement are allowed, with restrictions, and the push to quantify is always present. The 3Es logframe further constrains possibilities by forcing projects and programmes to figure a simplified linear relationship between costed inputs and quantified outcomes and impact. Values that cannot be quantified and thereby fitted into a clear linear relationship to inputs are framed out of the VfM equation, meaning they don't count in decision-making about what to fund and where. Accordingly, projects and programmes trying to 'sell' themselves to donors in the quasi-market described above will increasingly focus on projects that deliver narrowly defined VfM over other types of potential value.

Conclusion

Marcus Leroy (2011, no page), a veteran of the international aid community, has written critically about VfM, arguing that:

> As much as utilitarianism, 'value for money' is fundamentally consequentialist. That means acts, in this case activities of an aid donor, are judged by their results, not by any preset ethic rules. Indeed, the 'value for money' approach is inspired by the idea that we can get rid of the political or moral dimension of the issues involved. It wants to translate every objective into a 'deliverable', a rational objective that has nothing to do with ideology.

Of course VfM, and the aid effectiveness agenda of which it is a part, originated in the effort to make aid fairer, more accountable, and, of course, more *effective* for its beneficiaries as well as for donors. Indeed, replacing ideology with rational decision-making in development aid seemed a laudable idea in the wake of the excesses of the cold war. It would be easy to say, as Leroy implies, that the limitations of VfM are evidence that the pendulum has swung too far in the direction of rationality, creating an 'iron cage' for development (cf. Weber 1958). This chapter has presented a more nuanced critique of VfM, demonstrating that it is not necessarily rational. After all, it is not rational to assume all development outcomes that matter to beneficiaries and donors can be clearly quantified and related to costed inputs! Moreover, it is not at all clear, as Leroy suggests, that DfID or the consultants and contractors it works with believes VfM will dispense with the political or moral dimensions of decision-making in the allocation and use of development aid. Gut feelings and expert judgement are undeniably part and parcel of the calculations. Rather, this chapter has shown that VfM acts as a tool for *performing* rationality. VfM performs rational decision-making as a justification for continued spending on international aid at a time of domestic austerity; and, more ominously, it performatively shapes what counts as development value by constraining the possibilities for measuring and framing that value to fit its conception of rationality. Instead of swinging the pendulum back to ideology, then, we should start thinking about opening up the category of what counts as 'rational'. We might replace the limited and limiting practices that currently constitute VfM and aid effectiveness with practices that embrace political, moral, and contextual considerations.

Notes

1 Interview, London, 04/07/15.
2 For detailed histories of international development aid see, for example, Allen and Thomas (2000); Rist (2003).

References

Allen T. and Thomas A. (eds) (2000) *Poverty and development* Oxford University Press, Oxford.

Boin C. (2010) "Cutting waste at DfID has only just begun" *The Telegraph* www.telegraph.co.uk/news/politics/conservative/7735786/Cutting-waste-at-DFID-has-only-just-begun.html (Accessed 24 July 2017).

Bond (2011) "Value for money: what it means for UK NGOs: background paper" Bond, London, https://www.bond.org.uk/data/files/Value_for_money_-_what_it_means_for_NGOs_Jan_2012.pdf (Accessed 24 July 2017).

Butler J. (2010) "Performative agency" *Journal of Cultural Economy* 3: 147–161.

Çalışkan K. and Callon M. (2010) "Economization part 2: a research programme for the study of markets" *Economy and Society* 39: 1–32.

Callon M. (ed.) (1998) *The laws of the markets* Blackwell Publishers/Sociological Review, Oxford; Malden, MA.

Cochoy F. (2008) "Calculation, qualculation, calqulation: shopping cart arithmetic, equipped cognition and the clustered consumer" *Marketing Theory* 8: 15–44.
Davis K.E., Kingsbury B. and Merry S.E. (2010) "Indicators as a technology of global governance" *Law & Society Review* 46: 71–104.
Davis T.W.D. (2012) *ACFID and 'value for money'* Discussion paper prepared for ACFID. ACFID, Canberra, https://acfid.asn.au/sites/site.acfid/files/resource_document/ACFID-Value-for-Money-Discussion-Paper.pdf (Accessed 23 August 2017).
DfID (2011) *DfID's approach to value for money (VfM)*. DfID, London, http://mande.co.uk/blog/wp-content/uploads/2011/08/pub_031035.pdf (Accessed 23 August 2017).
Dixon L. (2010) "DfID accused of wasting money on taxis and offices" *The Times*, www.thetimes.co.uk/tto/news/politics/article2785004.ece (Accessed 24 July 2017).
Fourcade M. (2011) "Cents and sensibility: economic valuation and the nature of 'nature'" *American Journal of Sociology* 116: 1721–1777.
Fredriksen A. (2012) *Making humanitarian spaces global: coordinating crisis response through the cluster approach*. PhD Dissertation, Department of Sociology, Columbia University, New York.
Hansen H.K. and Porter T. (2012) "What do numbers do in transnational governance?" *International Political Sociology* 6: 409–426.
ICAI (2011) *ICAI's approach to effectiveness and value for money (No. 1)* Independent Commission for Aid Impact, London, https://www.bond.org.uk/data/files/ICAIs-Approach-to-Effectiveness-and-VFM.pdf (Accessed 23 August 2017).
Jackson P. (2012) *Value for money and international development: deconstructing myths to promote a more constructive discussion*. OECD Development Cooperation Directorate, www.oecd.org/development/effectiveness/49652541.pdf (Accessed 23 August 2017).
Krause M. (2014) *The good project: the field of humanitarian relief NGOs and the fragmentation of reason* Chicago University Press, Chicago, IL.
Leroy M. (2011) "'Value for money' or 'results obsession disorder'?" *The Broker*, www.thebrokeronline.eu/Blogs/Busan-High-Level-Forum/Value-for-money-or-Results-Obsession-Disorder (Accessed 24 July 2017).
MacKenzie D. (2003) *An engine, not a camera: how financial models shape markets* MIT Press, Cambridge, MA.
MacKenzie D., Muniesa F. and Siu L.(2007) *Do economists make markets? On the performativity of economics* Princeton University Press, Princeton, NJ.
Merry S.E. (2011) "Measuring the world: indicators, human rights, and global governance" *Current Anthropology* 52: S83–S95.
Mitchell T. (2002) *Rule of experts: Egypt, techno-politics, modernity* University of California Press, Berkeley.
Mosse D. (2004) "Is good policy unimplementable? Reflections on the ethnography of aid policy and practice" *Development and Change* 35: 639–671.
Mosse D. (ed.) (2011) *Adventures in Aidland: the anthropology of professionals in international developmen*t Berghahn Books, New York.
Muniesa F. (2014) *The provoked economy: economic reality and the performative turn* Routledge, London.
Muniesa F. and Linhardt D. (2011) "Trials of explicitness in the implementation of public management reform" *Critical Perspectives on Accounting* 22: 550–566.
OECD (2017) *Paris declaration and Accra agenda for action*, www.oecd.org/dac/effectiveness/parisdeclarationandaccraagendaforaction.htm (Accessed 24 July 2017).

Poovey M. (1998) *History of the modern fact: problems of knowledge in the sciences of wealth and society* Chicago University Press, Chicago, IL.

Porter T. (1995) *Trust in numbers: the pursuit of objectivity in science and public life* Princeton University Press, Princeton, NJ.

Rist G. (2003) *The history of development: from Western origins to global faith* Zed Books, London.

Scott Z., McLoughlin C. and Marquette H. (2012) "Combining value for money with increased aid to fragile states: welcome partnership or clash of agendas?" *Crime, Law and Social Change* 58: 509–519.

TNS UK (2010) *Public attitudes towards development.* TNS Report prepared for the Central Office of Information (COI) on behalf of the Department for International Development. TNS UK, London, https://www.gov.uk/government/uploads/system/uploads/attachment_data/file/67684/public-attitudes-april10.pdf (Accessed 23 August 2017).

Vardakoulias O. (2013) *New economics for: value for money in international development* NEF Consulting, London.

Weber M. (1958) *The Protestant ethic and the spirit of capitalism* (trans. Parsons T.) Charles Scribner's Sons, New York.

4 The *value* of human life in health systems and social spaces

The HIV/AIDS context in Zimbabwe

Fortunate Machingura

Introduction

In this age of increased wealth, advancements in medical treatments and globalisation, the gap between richer and poorer social groups has not only persisted but, in many cases, has widened. New research published by economists from the Universities of Pennsylvania and Michigan (Sacks et al. 2012) shows that as income rises, the well-being of an individual, community or country also improves and that richer social groups report higher well-being than poorer populations. That such a divide exists is unquestionable, however the dynamics of this divide and its consequences for societal perceptions of individuals' *value* have not been fully explored. For example, could it be said with certainty that wealthier people are treated better and valued more in society than their poorer counterparts? Drawing on two excerpts from life history interviews administered in an extended ethnographic enquiry in Zimbabwe with people living with HIV/AIDS, this chapter explores how the value of human life is framed in health, with the aim of identifying societal perceptions on the worth or value of human life.

The narratives speak to a broader, vernacular problematisation of value, in which significant gaps between the haves and the have-nots exist and a moneyed, powerful elite lives longer and happier. This, while the much larger majority scrapes a precarious living from health systems threatened by a falling human resource base, budgetary cuts and shortages of medication. While this chapter is not an exhaustive examination of every possible problem in the *value-ation* of human life, it aims to identify local epistemologies of value and bring together the interconnections of power, difference and resistance. It demonstrates not only how senior government decision-makers use calculative rationalities as critical *value-ation* tools, but also how they develop a peculiarly open numerical space which serves as an alternative to rationalising choices on who lives and who dies. A more flexible, vernacular vocabulary of value that translates across a broad spectrum of discursive spaces enables, in turn, a more accessible reading of the ways in which human life can be valued in health systems and social spaces, through a careful (non-totalising) interpretation of governmentality. The chapter concludes by considering a reading of the concept of *allowable death* (justified through consciously crafted narratives around the inevitability of death), as death that is both premature and avoidable. First, however, the chapter begins by clarifying how Zimbabwean vocabulary frames the 'value' of human life.

The problematisation of the value of human life in vernacular theory

The concept of *value* has obscure etymological origins. In English, the term is rooted in metaphors of value as wealth and as health. The etymology of the word can be traced to the 13th century. It is derived from the French noun *valeur,* and the Latin verb *valere,* meaning 'worth' and 'to be in good health' respectively (OED 2003). Value is also understood to correspond with notions of worth or the regard in which something is held (OED 2003). As such, value is a homonym, with more than one meaning, going beyond its restrictive etymological dichotomisation. This homonymity is also evident in African conceptualisations of the term, where spatial metaphors for value and valuation create multiple meanings (Machingura 2016, 20). Here, value is conceptualised as a noun, and as a practice (of value-assigning, value production, or both). For instance, in Ndebele, spoken in Zimbabwe, South Africa and parts of Namibia, value is *ukuqakathekisa* (meaning, to make significant) and *ukuhlonipha* (to respect, to honour). In Shona, spoken in Zimbabwe and parts of Mozambique, value is metaphorically understood as *hukoshi* (importance), *huremu* (worth), *mwero,* (standards/principles), *mutengo* (pricing), *rukudzo, chiremera* (respect, worth, honour) and *hunhu* (personhood). In Chichewa, spoken in Malawi, Zambia and parts of Zimbabwe, *ulemu* (respect), *ulemelero* (worth) and *malinzi* (cost) are expressions which may also signify value (Machingura 2016, 20–21).

Thus the concept of value carries various forms of meaning in nuanced variations of language. It is a multi-attributed concept which may focus on implicit ways of valuing human life. For instance, it can include the words that people say and how they are articulated; the rationale behind the words and context in which they are said; as well as the audience to whom the words are directed. All our perceptions about the world around us, our conceptualisation of the self, our relations with others, along with our feelings about the singular every-day, emerge from a combination of social factors. These factors also establish socially constructed nuances of the ways in which human life is valued (Machingura 2016). Thus the value of human life emerges as a social construct. As a result of these social assumptions and definitions, people can be assigned categories of value through the enactment of taken-for-granted rules (e.g. laws, policies, resource allocation formulas). Together, these assumptions and sense-making shape how people are valued in society, along with the way others define their worth based on material possessions or entitlements – a local construction of value.

However, this local construction of *value* is not represented in mainstream development and scholarly debates. Instead, the concept of value is often presented as 'stipulative', implying that only ways and methods that meet certain universally accepted criteria can be used to describe the valuation of human life. For instance, civil courts tend to consider the value of human life when awarding compensation to the families of deceased victims of malice or negligence. They also assign the value of human life in vaccine injury/death compensations, workers' compensation benefits and military death gratuities (Friedman 1982; CDC 2011). However, these valuations are not limited to civil courts alone. Public policies also involve consideration of similar issues in decision-making about resource allocation, where benefits and costs need to be expressed in one standard unit (Mishan and Quah 2007). This enables comparison of the relative value of saving different groups of human lives, with achieving competing policy objectives. This rationale becomes the primary logic for assigning dollar values to human lives (Mishan and Quah 2007). Relatedly, there is a consensus in the human rights literature that human life carries an intrinsic value (see UN 1945, 1948; Feldman 2002; Alberti and Emmons 2017), which does not equate to the monetary sums emerging from compensation and insurance claims, or in calculations of policy impact.

On human rights and the value of human life

Providing the strongest support for the claim that human life has an inherent value, the UN Universal Declaration of Human Rights (UDHR) adopted by the UN General Assembly in 1948, (UN 1948) stands above other texts such as the Bible or Koran. The UN UDHR provides a compelling justification for the proposition that, because human life matters, it has a normative intrinsic value which is universally known and upheld (Alston 1999). Following a universal recognition through the UDHR that human life has inherent value, the adoption of the International Covenant on Civil and Political Rights (ICCPR) in 1966, paved the way for a universal commitment to observe and respect that every person holds the right to life (UNHR 1966a). Also, the International Covenant on Economic, Social and Cultural Rights (ICESCR) of the same year (1966), contextualised this in the form of rights to cultural expression, dignity and social autonomy. The ICESCR, in particular, espouses that for a human being to live with dignity, they must enjoy their right to health. The right to health is linked to the realisation of other human rights, including rights to life, food, education, dignity and equality (UNHR 1966b). Both the ICCPR and the ICESCR affirm that the right to life comes from the intrinsic value of human life, and not necessarily from legal provisions that countries offer. Correspondingly, an international 'worldview' understanding asserts that the life of every human being is equally worthwhile, regardless of any differences (see Amar 1998; Alston 1999). Corroborating the assumption that human life has an inherent value are the works of Ronald Dworkin which argue strongly that it is the life of a person, and not necessarily the person's living body cells, that has 'inherent value' (Dworkin 1978). It is from this understanding of life (outside the living cells) that legal and ethical arguments on the termination of life, including abortions (Belshaw 1997), euthanasia (Machingura 2016), human clinical experiments, judicial murder and capital punishment (Slovic 2010), are drawn. Thus, it is reasonable to argue for the preservation of life and prevention of premature death at all costs.

It is within the concept of human rights that governments express their obligation to protect and prolong life and prevent death, through the application of international conventions, national constitutions and domestic laws and policies (UNHR 1966b; Ssenyonjo 2008). Similarly, prohibition of violent deaths and deaths without dignity, and averting preventable premature mortality are policies shared among nations (UN 1948). This rights discourse thus becomes the benchmark for asserting how the denial of rights arises, and how discrimination is built as the contra-condition to a life of rights and dignity. Correspondingly the literature on health rights and specifically people living with HIV/AIDS references the rights discourse in explaining discrimination. In particular, Crawford (1994), Parker and Aggleton (2003), Campbell et al. (2005) and Castro and Farmer (2005) discuss discrimination as an element of negative identity associated with people living with HIV/AIDS and arising from their involvement in social actions and relationships that devalue them. Discrimination amounts to a failure to respect human dignity by devaluing those affected, often adding to health inequalities (Parker and Aggleton 2003; Goudge et al. 2009; Machingura 2016).

Although there is a consensus in the human rights literature that human life has an 'inherent unique value', there is no conclusive empirical evidence to demonstrate that people recognise it in everyday practice and especially in health care. Instead, there are two major schools of thought regarding life and death in the health care arena. The first comes from medicine, which argues that public health has transformed death from a fate into something which could be viewed as a 'risk', managed through planning (albeit within limits because death is inevitable) (Rothman et al. 1995; Porter 1999; Bynum 2008). This school of thought is consistent with human rights perspectives on the value of human life – that human life has inherent value, which should be protected through policy and government interventions. The second school of thought comes from health economics (Culyer 1989), which argues that valuation of health care through evaluation of costs, benefits and other mathematical approaches is the best way to inform decision-making, including decisions on the rationing of health care resources. However, rationing comes with trade-offs that senior government officials, donors and others must manage within rule-bound systems of principles and choices, which ideally are transparent and viewed as fair by health care users.

Micro-rationing in economic health evaluation

Klein et al. (1996) divide rationing into two: micro- and macro-rationing. Micro-rationing is concerned with the rationing of care at an individual level (known as the second level of resource allocation) which occurs, for example, when health workers make decisions about which resources to allocate to individual patients. In micro-rationing, health workers use their judgements to evaluate the patient's clinical needs, which can include technical assessments to determine the seriousness of a particular disease. Hughes and Griffiths (1997), however, argue that such clinical evaluations can extend to social or moral assessments, such as when health workers ration the time they are available to provide health care services. This type of rationing becomes more personal than scientific. Personal rationing may include the ability of health workers to function competently and to offer quality services (Hughes and Griffiths 1997). This kind of personal rationing appears to be so embedded in the everyday use of 'scientific based (economic) rationing' that it may be difficult to identify and quantify. It could be that personal rationing, as argued by Hughes and Griffiths (1997), signals that an individual is suitable (i.e. deserving and worthwhile) or unsuitable (i.e. undeserving and unworthy) for care. Personal rationing practice as micro-rationing is predominantly an action of using 'common sense' to make rational decisions, or at least the common sense that each health care worker believes they are employing (Scheunemann and White 2011; Machingura 2016). This is significant to the question of how human life is valued, because micro-rationing decisions are thus essentially subjective and can differ from one health carer to another. This can result in varying outcomes, related, for example, to decisions about the rationing of in-patient care bed units, to medicines or which patient to serve first from a queue in the emergency department (Machingura 2016).

Macro-rationing in economic health evaluation

Macro-rationing is concerned with the setting of priorities about the kind of care that will be provided and where (known as the first level of resource allocation). It occurs at the policy level of the health system. Here, health valuation techniques or macro-rationing methods are used to value health as a proxy for valuing human lives, based on the costs of health care (Klein et al. 1996). The techniques that have been discussed most in the literature are found in the works of Dolan (2008) and Holmes (2013). Across this body of work, three major macro-rationing techniques are identified:

- cost-identification analysis (CIA), which considers the cost-per-individual intervention, but excludes costings of benefits or health outcomes;
- cost-effectiveness analysis (CEA), which tests whether health outcomes are the same for two interventions. If intervention A is cheaper and produces more efficient outcomes than intervention B, it prevails. On the other hand, if the first intervention is costlier but more effective, the extra benefit may nonetheless justify the extra cost. However, CEA only compares interventions whose benefits can be measured in identical units of effectiveness. As such, it fails to actively inform spending decisions on wider issues;

- cost–benefit analysis (CBA), which is used to inform decisions about how much to spend on wider issues such as livelihoods, protection or food in relation to health care or HIV and AIDS care. However, in this type of analysis all effects (both benefits and costs) are aggregated into dollar amounts. This can be controversial because it involves putting a monetary value on human life or asking patients what they wish to pay to avoid death from AIDS. It also fails to differentiate cost and outcome.

There is, of course, nothing fundamentally wrong with these economic or human rights constructions of the valuation of human life. The development of concepts, like all scholarship, must start somewhere. This contemporary understanding of the value of human life seems to be based, among other things, on what we know of, and refer to as, the fundamental right to life, worth or dignity. However, this does not mean that all people, at all times, have shared the same understanding of and linguistic terminology to describe the 'value of human life'.

Subjugated knowledges

The Zimbabwean construction of the concept of *value*, as cited earlier in this chapter, could be viewed in terms of what Michel Foucault calls *subjugated knowledges*. Subjugated knowledges allude to bodies of knowledge that are regarded as inferior to the established regimes of thought or "beneath the required level of erudition or scientificity" (Foucault 1980, 81–82). Foucault defines these as 'low-ranking knowledges' that are not subjected to centralised spaces of theoretical production; the knowledges whose validity is not dependent on the approval of the established "regimes of thought" (Foucault 1980, 81). The many examples of meanings attached to the concept of value in the Zimbabwean context alone illustrate very well the strikingly different forms that the value of human life is perceived to have in different contexts, as well as the various concepts of value that people hold. This last point is crucial. To these contextual examples that define 'value', one could add a myriad of anthropological case studies, which indicate that ideas, of what is close enough to be described as the value or worth of a person, are nevertheless different in striking and important ways. Here an attempt is made to bring together the local or vernacularised theories of *value,* with the health sector's ideas of valuing health as proxies for valuing human life. Drawing from ethnographic enquiry conducted over several months in Zimbabwe, two stories of people living with HIV/AIDS are contrasted to examine the ways in which health systems and society frame the value of human life and, specifically, people living with HIV/AIDS. Subjecting two cases of people infected with HIV to value theories, provides an opportunity to understand how society weighs the value of life between those who are 'deserving' of sympathy, help, support and treatment, and those who are not. These theories may lack the intellectual weight of academic theory, but I propose that class prejudices and elitism in academia blind us to the theoretical contributions that these knowledges can make, particularly when they question whether all lives do indeed claim equal value.

The ethnographic cases: the haves and the have-nots

Selected from over 110 interviews with people living with HIV/AIDS, what follows are two distinct but overlapping ethnographic narratives about the value of human life that we seldom hear/encounter, yet they contradict what we believe to be stable in the world – that every person has an inherent value.

Value and social status – the case of Ncube

Dr Ncube [actual name withheld] is a former government minister, senior official and member of parliament, as well as being one of the country's most venerated politicians. In 2013, he was involved in a car accident on one of Harare's busiest roads and broke his neck. He was rushed to Parirenyatwa Group of Hospitals, where upon arrival, in pain and distress, he immediately revealed his identity to hospital staff. As he recounted in an interview with me in 2014 "the health system still works efficiently, in a few minutes of my arrival at the hospital, it seemed, my cubicle was filled with specialist doctors, emergency personnel, nurses, nurse aids and other hospital staff".

His X-rays and magnetic resonance imaging (MRI) scan showed fractures of two cervical vertebrae and multiple fractures to the skull and face, with an acute subdural haematoma from trauma associated with the accident's considerable impact. Emergency medical personnel, doctors and nurses reacted quickly and with urgency. They administered large doses of steroids, intravenous liquids and vitamins, all carefully calibrated to take into account his regular medications. However, due to airway obstruction resulting from substantial haemorrhaging, Ncube began choking. The hospital's attending senior anaesthetist tried in vain to insert a breathing tube into Ncube's trachea. A physician from the nearby trauma centre in Avondale, about half a mile from Parirenyatwa hospital in the capital city, was immediately summoned and, upon arrival, performed an emergency tracheotomy, which enabled Ncubeto breathe again.

However, soon after, he had a sudden cardiac arrest. The emergency medical personnel responded immediately with cardiopulmonary resuscitation and injections to stimulate his heartbeat. Within two minutes his heart had re-started, but stopped twice more in the ensuing 20 minutes. As Ncube described:

> My cardiologist, from Harare's prime trauma centre, a senior and venerated physician, arrived in the emergency room at Parirenyatwa in the nick of time. He quickly advised the entire medical crew on how

> to manage a sudden cardiac arrest while also managing my HIV status along with my unstable heart rhythms and blood pressure. I thought I was dying, but his expertise and the cooperation and efficiency of everyone in that room on that day saved my life.

Ncube was hooked to a ventilator, a Foley catheter was inserted into his bladder to drain urine and a bronchoscopy was administered to evacuate blood clots from his chest.

Soon after, Ncube was taken to the Intensive Care Unit (ICU), where, he stressed, his wife, children and grandchildren gathered at around 2 am. Other senior politicians, his GP (who had been called to give information on Ncube's HIV drug regimen), four lawyers and three army generals were also present – a gathering that in other circumstances would not have been permitted. This case however, was different:

> The medical team kept informing my family about my conditions and the procedures and were very polite and answered all their questions. My fellow politicians were offered tea and constantly told that it would be OK. ... the hospital staff did everything they could to relieve my discomfort. To be honest, they were always available day and night, competent, quick, clean, polite and incredibly loving.

Ncube spent 20 days in the ICU before being transferred to Dandaro Clinic, located in Borrowdale, one of Zimbabwe's most affluent neighbourhoods in the capital city, where he received specially tailored rehabilitation services and continued his recovery for another three months. The total cost of his medical care, with more than half covered by his medical aid insurance and the remainder paid by his employer, came to USD 250,000.

Clearly, Ncube's economic and social status derived from the honour and prestige attached to his position in society and the value of life that he could claim from such a standing was high. His elevated status was reflected in how the health system responded to his needs, demonstrating how privilege is built into the system, through various forms of institutionalised behaviour and other arrangements. The value of Ncube's life was distinguished and accorded a differential status in line with society's framing of worth, value and dignity. Moreover, at the time of his injury, Ncube was 70 years old, in a country where life expectancy is below the age of 50 years.

We are worthless, like old clay pots that have lost their value – Tino's household

We agreed to meet at Charity's rural home, where she now lived with her husband and their three children. Waiting in the distance in the scorching sun, through bleating goats and clinking cowbells, stood a frail-looking figure. Although bent over and clutching a thin walking stick, I could tell from the happy waving and beckoning that this was Charity. As I approached, she smiled and waved, showing the way. Using her walking stick, she pointed in the direction of a man lying on his back, whom I guessed was Tino, her husband.

Charity, aged 33 years, was living with HIV. Her husband Tino, 37 years old and also living with HIV, was now back home after having been incarcerated for 10 years. Although they lived together, they barely talked, being afraid of upsetting each other. Thus, they mostly avoided discussing their illness at all. Charity directed me to a reed mat, and gently put her hands under Tino's back to help him into a sitting position on the edge of the wall. Emaciated and silent, Tino sat and leaned on the soft pillows; he tried to pull up his trousers, now loose and too large for him, with no belt to secure them. Neither spoke for several moments. Then, as Tino broke the silence, Charity interrupted, calling for a cup of water from her eldest daughter, Rudo (16 years old), who had just arrived along with her 2-year-old brother Tonderai on her back.

Rudo spent most of her time caring for the home, her parents and young brother. Today, she had penned the cattle on the mountain side and had carried out most of her morning chores. Rudo had gone to fetch water from the communal borehole, 20 minutes' walk from their homestead, with Tonderai on her back. She left her 20-litre water bottle on the water queue while her sack of maize flour was being ground for the weeks ahead. Rudo returned to pick up her full water bottle and carried it home, along with a bag of corn flour on her head, and Tonderai on her back. As she brought water to us, she told her mother that she would soon leave for the fields to weed sorghum with a group of 10 people, who had arranged to work in each other's fields in reciprocal exchange. Rudo carried her baby brother on her back during the day, interrupting her work to feed the child when he cried. She fed him with maize flour and millet beer when the workers stopped for lunch. Watching from the homestead, I saw Rudo, along with others, working hard, as a line of men hoed ahead, and a line of women progressed steadily through the terraced fields.

Charity was proud of her hard-working daughter: "she takes care of us, I feel sorry for her, but there is nothing much I can do these days, I feel weak". Charity's HIV diagnosis was made in 2011 while she was pregnant with Tonderai, three years after Tino returned from prison. Tino knew about his HIV status when he was in prison, and blamed prison staff for failing to provide condoms to inmates regularly. Even more significantly, however, Charity explains how HIV, along with Tino's hidden sexuality "cheapened" their social status and the way in which they were viewed by the community:

> Yes, we both have HIV, but I am on ARVs [antiretrovirals] and Tonderai is negative thanks to the ARVs. Tino won't go back to the clinic because the nurses are very unfriendly to him and the whole community knows 'nezve hungochani hwake' [that he is a gay man]. The last time we went to the clinic together the nurse was extremely rude, almost angry at him. People laughed at him, and it made me angry. I'm also mad at Tino because he did not tell me the truth about his sexuality!

As she narrated her experience, she displayed a range of negative emotions, from extreme depression, through to anxiety, anger and hopelessness. Charity spent most of her days thinking about the negative judgements she perceived others held of her, her children and her extended family.

The unexpected shattering of trust (and with it the life she thought she had), coupled with HIV, her husband's homosexuality, their poverty and lack of income, compounded Charity's sense of shame and devastation. She explained how for years she denied her pain, leading to what seemed like permanent agony in the form of a debilitating self-image – a profound worthlessness:

> Vanhu vanotiona semarara, sekunge mbwa chaidzo, se tinosemesa nemhaka yezvakadai izvi. Hatichisina kana chiremera, tinongovewo hari dzofanzirofa [people see us as dirt/chaff; as if we were like dogs like we disgust them because of this situation you see here. We are worthless, like old clay pots that have lost their value in the kitchen].

"It didn't take long to get through the shame and mental thing" Charity says, with what would emerge as characteristic understatement:

> The first few months were horrible because I only got to know about my status when I was pregnant with my youngest child – I've always had an inclination towards suicide ever since my husband left me with our first two kids when he got into prison. When he came back home, I was fighting to get food in the home. Tino was getting worse by the day, he wasn't working, I was doing everything – the home, the children, we were homeless in Harare, and that's why I decided to move back to our rural home.

Charity was also fighting to come to terms with her diagnosis. Tino, who had been quiet all along, began to speak:

> I know I wronged her, and we have discussed this several times, but again and again she continues to bring back this conversation with such anger and rage. If anything, I'm the one on the sick bed, and I am like a fly caught on a window pane, bending itself this way and that, buzzing wildly and dying stupidly, without dignity, with no one and with nothing.

Anger ran through their coping: a mixture of rage at the deception and grief arising from awareness of the loss of all they had: their belief system and the esteem of others. They were in despair.

Like Tino, Charity's mental and physical health worsened with time. She described the times she had to deal with depression along with the recurrent fungal skin infections that are common among immune-suppressed individuals. The memory of her experiences compounded her feelings of worthlessness.

By May 2011, Tino's CD4 count – the primary measure of the immune system for HIV-positive people – had dropped sharply to 112, the point at which doctors confirm an AIDS (late-stage HIV infection) diagnosis. Although Tino knew that taking ARVS would help him, he had continuously refused treatment (the subconscious reason for his resistance only surfacing many months later). Finally, however, he agreed to start combination therapy, but the clinic was too far from his rural home and upon finally reaching it, the nurse there was unfriendly. Soon, he developed pneumonia, more common in patients in the advanced stages of HIV when the CD4 count is low, than in the general population. However, within a month this was successfully treated.

Depression nonetheless, took hold again:

> I thought to myself, if I live, that's good, but if I don't, well that's life, after all, everyone dies eventually. I don't think my wife and family realised how bad my mental state was.

Tino still had a persistent cough, and his wife feared it might be tuberculosis (TB). She feared that Tino would die soon and that she along with her children might also get TB. However, Tino vowed not to go back to the clinic:

> They don't care about us, if you are poor like me, it's worse, it depends on wealth, and the respect that you have. Even though the law says no to discrimination, it doesn't help because we deal with real people who have beliefs and they will judge you on these terms.

The homophobic abuse from medical staff and the general community that he experienced made him think that he was already dead: "It's as though I was already dead, it's a nightmare just to take a step" he said sighing and pointing at his swollen feet. There have been periods when, so overwhelmed by his health problems, it is all he could talk about with his three children. Relatives did come to see him but not as often these days:

> Relationships with friends or my extended family is only a memory; it's this HIV situation plus my sexuality. People refuse association with these two things.

For Tino, the latent low self-esteem, seeded in the community and a health system plagued by homophobic abuse, was compounded. With a deep sigh he said:

> I thank God that I'm still alive and I know my wife will live longer than me to see Tonderai grow. She is a fighter, and despite our arguments, she has remained by my side. She has prayed for me and got me herbs to apply on my wounds.

Despite the despair, mechanisms of belonging, value, tenacity and family care were still worth attending to.

In relation to the opinion that HIV is no longer a death sentence because treatments have converted HIV to a chronic lifelong disease, Tino took a different view:

> That's nonsense, it means different things to different people, to me and others like me; it means losing everything including one's identity as a person. The wealthier can surely buy long life.

He spoke calmly, with head slightly bowed, but maintained constant eye contact with me.

Tino's case is emblematic of the perils inherent in interactions between HIV infection, poverty, sexuality, access to treatment and the 'constellated' disadvantage that some people living with HIV/AIDS face, in particular the hardship of how social worth interacts with the value of human life in Zimbabwe.

Discussion

The experiences of Dr Ncube, Charity and Tino here uncover a variety of issues which, while not often seen together in one household, are separately found in many. Their characteristics seem all too familiar and are rooted in the hard realities of sub-Saharan African experiences and the contexts of most developing countries. These narratives encapsulate a broader vernacular problematisation of value, significant gaps between the haves and the have-nots – where moneyed and powerful elites can live longer and happier, while the majority scrape a precarious living from health systems threatened by a falling human resource base, budgetary cuts and shortages of medication.

People like Ncube, Tino and Charity who are living with HIV/AIDS and getting older, need – and deserve – more from health services. Advances in technology and improvements in the effectiveness of ARV therapy and expertise in medically managing HIV have helped to increase the chances of survival and improved the life expectancy of people living with HIV/AIDS. But all these developments, and especially in Zimbabwe, result in an ever-increasing health care expenditure in an already weak economy. By the end of 2015, the Zimbabwe Demographic and Health Survey reported an HIV prevalence of 15 per cent, with more and more people still needing lifesaving ARV treatment (ZIMSTAT and ICF International 2015; MoHCC 2016). This context begs several questions: how far can the health delivery system go to save Ncube's and Tino's lives? What is the value of their lives? Should we be prepared to save both, including Charity, at all costs? Or should we think about the financial consequences of our decisions? What are the implications for the lives of these individuals?

On the surface, these may be relatively easy questions for a health economist whose task is focused on assessing how health expenditures can sustain a functional health delivery system. For instance, the Zimbabwe AIDS levy established in 1999 by an Act of Parliament has generated considerable resources, with recent estimates averaging USD 35 million per year (Bhat et al. 2016). This continued increase in the AIDS levy revenue over the years signals a significant commitment by Zimbabweans to prevent HIV/AIDS and purchase ARVs for treatment, support and care (Bhat et al. 2016). The AIDS levy entails a 3per cent income tax for individuals and 3per cent tax on profits of employers and trusts. However, even for the seasoned economist, decisions on which lives should be saved, where, and whether age should be an influencing factor, do not come easy. Is there an assumed or calculated depreciation in value over time (especially when people get infected with HIV), or is there a value appreciation in the context of improved education, affluence, prominence and total contributions that we make to our communities, business, friends and family?

If *value* is merely a matter of deciding who is more important, then we may simply turn to utilitarianism and economics. In this case, one could justifiably assign importance to a baby with a whole life of work and productivity ahead, over a dying geriatric. For Tino, in his thirties, could we argue that the health system should have been more responsive than for Ncube? If *value* is a matter of elitism, and the tax contributions that a person can pay, it is not so simple. In this reading, wealthier, employed and tax-paying members of society such as Ncube are valued more than poorer, unemployed members such as Tino and Charity.

In the Zimbabwean context of the AIDS levy, the economic prospects of Ncube and Tino are pertinent. An economic calculation alone would suggest that it is not in the interests of a society to pay a high price to save Tino's life while his employment chances or capacity to employ others are low, and the expense cannot be recovered from him. From a purely utilitarian viewpoint, this appears to be an unprofitable expense. Admittedly, it could be 'fair' if the person who incurs the medical bill has duly paid taxes all his/her life to cover the medical costs of others. However, this does not change the fact that in the case of knowing a patient is of no further economic value, it would be unprofitable to meet his/her treatment costs. A different option to address this seemingly unfair attitude would be to shift the responsibility to meet medical expenditures to personal financial resources. For example, Ncube could pay his own bill. However, issues of equity are not eliminated here. Ncube's ability to pay is related to another's' inability to pay, while his use of medical resources prevents them being distributed to others. Meanwhile, Tino and Charity cannot cover their medical expenses at all, such that structural poverty assigns to them an early death in the absence of a redistributive mechanism. Although both Tino and Charity could access free ARVs from their local clinic, discrimination drives Tino away. He would have continued ARV treatment only if there were options for private health care that he could afford. However, it would seem unfair to let Tino die because he doesn't earn as much money as Ncube.

The implicit value of a life saved

The government of Zimbabwe makes a number of spending decisions which indicate how willing it is to reduce loss of life. Examining these decisions can give us another perspective on the cost-per-life saved. In measuring costs, health economists and researchers in the Ministry of Health and Childcare (MoHCC) first identify HIV/AIDS resources that may need to be used, then quantify and finally assign a monetary value to such resources. The calculations are administered to compare the costs of interventions per-every-health-outcome attained in relation to social objectives, i.e. maximising population health (and, in this case, the health of people living with HIV/AIDS). Recent cost-effectiveness analysis studies of hospitalisation and home-based care for people living with HIV/AIDS in Zimbabwe (Hove-Musekwa et al. 2014) suggest that in these calculations, no monetary value is assigned to outcomes, but rather results are presented in the form of cost-per-every-health-outcome. Results are shown as costs (direct, indirect and intangible) per HIV infection averted, and cost-per-life saved. Indirect and intangible costs, such as the pain of suffering (e.g. morbidity), are those which are difficult to measure because they do not have a market value. The CEA calculation considers both morbidity and mortality (death) costs. Ultimately, however imperfectly, the economic impact captures both the cost and cost-effectiveness of the various HIV/AIDS and HIV-related interventions. An intervention's effectiveness is measured in terms of health outcomes (health benefits) such as quality-adjusted life years (QALYs) or merely infections averted. Such a measurement reflects the valuation that one year of life living with HIV is less desirable than one year of life without HIV, and a year of life with asymptomatic HIV infection is more desirable than a year of life with symptomatic HIV infection (Edwards et al. 1998).

Therefore, individuals like Tino and Charity who move from a health care environment in a city where transport services enable better access to health care than in rural areas, experience reduced health benefits as a result. Health benefits are measured by the duration of life years gained as a result of an intervention. Decisions to implement a particular intervention do not only depend on CEA, but also on factors such as what policy-makers in the MoHCC and donors consider as good value for money and are willing to pay for. This is variously informed by CEA calculations. Thus, while in theory a year of human life should be priceless, in reality, the cost-per-life value through CEA produces a monetary figure that donors, health insurance plans and policy decision-makers all consider when making decisions. Such a move could mean that if the cost of new technology or treatment is more than the monetary threshold, insurers, government administrators and politicians alike may recommend that treatment is not covered, because of the unnecessary risk and short-term benefits for quality-of-life.

Assigning a dollar figure to people living with HIV may sound crass, but such valuations are routine in Zimbabweans' daily lives. Take, for example, the cost calculations that for-profit medical aid companies make to determine how much coverage they give to their clients – they value healthier employed citizens relative to terminally ill unemployed citizens. A cost–benefit analysis model can deny life-saving treatment, well-trained health workers and medical technologies to the poor in mostly rural areas, because the clientele therein are 'worth-less' according to cost-effective health delivery formulae, when compared to their urban counterparts. Preventable premature mortality associated with such unavailability or shortage of health resources (being deemed not cost-effective), is what I term *allowable deaths*. Allowable deaths emerge as premature, avoidable deaths that have been left to occur under preventable conditions (i.e. conditions that are amenable to change). In these cases, however, there is a conviction on the part of the organisation (be it government, community, clinic or household), that the death was inevitable because nothing, under the prevailing conditions, could have been done to prevent it. The reality is that these deaths are unnecessary, unjustified, premature and preventable. However, large insurance companies that employ CBA calculations in policy make claim to justificatory economics on these issues.

'Hari dzofanzirofa' as value

Hari dzofanzirofa [chipped, cracked or otherwise broken clay pots of little, if any, use] is a Shona metaphor used by Charity to describe how people value her and Tino in society. The term relates directly to the quality-of-life of geriatric or critically ill individuals. Using cracked clay pots to metaphorically represent one's quality-of-life provides the cognitive framework for our perception of their reality. Here, Charity intentionally uses the clay pot metaphor in a specific discourse to convey a specific message. Her choice of *hari* (i.e. clay pots) reflects her idea of the current quality-of-life that she and Tino had, while simultaneously emphasising the imminence of death (*kufanzirofa*) and the fading away of life. From this perspective, quality-of-life refers to the moral worth or value of a person and his or her life. The notion of having a good quality-of-life emerges as part of a broader normalising ideology in mainstream culture. This could be perceived as one that works to devalue the lives of persons living with HIV/AIDS or other chronic conditions, by perpetuating stigmatisation and discrimination against them. The view of the quality-of-life as *hari dzofanzirofa* challenges the inherent dignity of the human person. To say that Tino had no quality-of-life or had a low quality-of-life, suggests that prolonging his life has no moral significance, either to Tino himself, to his wife and children or to society in general. Tino could simply be waiting for death and Charity wishing Tino to die in the shortest possible time – because they have given up. However, pushed to its logical extreme, this usage of '*hari dzofanzirofa*' to represent quality-of-life takes us to the ill-reputed Nazi concept of "life unworthy of life" (*Lebensunwertes Leben*). The Lebensunwertes Leben was a rationalising strategy to devalue human life through active euthanasia of those living with disabilities in the genocidal death camps (Eibach 1998).

The notion of the value of human life is logically quite distinct from the notion of quality-of-life that Charity describes, and is again different from the technology of calculation used by health professionals to 'make' quality-of-life. An account of the value of human life based on Charity's usage is grounded in an underlying account of respect, humanness or the human person; an account, that is, of what it is to be (*munhu*) human. On the other hand, the concept of quality-of-life presented as QUALYs gained (as described earlier) is based on calculations of a person's capacity to become (more fully) human, rather than being embedded in the intrinsic value of being human itself.

Religion as a double-edged sword – valuing and devaluing lives

Unlike Ncube who stated that the health system saved his life, Charity and Tino did not benefit in the same way. Instead, Charity and her husband Tino incorporate spirituality (God and, specifically, Christian doctrines) as a way to cope, reframe their value and bring a sense of meaning to their lives in the face of HIV. Such use of religion in daily life is associated with improvements in health-related quality-of-life and overall well-being. Findings here echo those from earlier studies showing that greater levels of spirituality could be associated with resilience, a greater sense of purpose to live and better overall health-related quality-of-life in people with HIV/AIDS. As such, religion can function as a source and space through which resilience as human agency is manifest – where the person's quality-of-life is heightened socially through belief in a transcendental being who watches over them and provides the armoury to fight physical and mental problems (Coleman and Holzemer 1999; Coleman 2017). By the same token, there are also harmful effects of religion for people living with HIV/AIDS, and especially homosexuals, who may be ostracised due to lifestyle issues associated with being HIV-positive. What is interesting here is how coping through prayer and God is, in itself, a mechanism of self-value where people like Tino and Charity get a sense of self-worth, family support and sympathy from others. Equally interesting however is how local, moral norms that are founded on the same religious doctrines also work as a means to exclude, stigmatise and bully gay individuals, or those perceived to be gay such as Tino. Thus, religious influence is double-edged, it can value human life as a source of peace, love, inclusivity and support, but it can also work like a brute coercive force in condemnation of homosexuality – undermining and devaluing life while advocating violence in its name.

The workings of power and value-assigning

In the ethnographic case study of Tino, religious influence and the perceptions of health workers and those around them emerge as critical nodes for the workings of power over people's consciousness. In Foucault's dealings with power, health institutions and religious organisations deploy the hegemonic power of language that is located in the discursive space to frame societal moral conduct (Foucault 1977). Further, the clinics and churches that control behavioural actions – like Foucault's asylums or prisons (as disciplinary institutions) – must thus be seen as deploying strategies of power and domination (Foucault 1982). Foucault characterised the subjection of the self to these many, pervasive mechanisms of control as disciplinary normalisation. Such modern disciplinary normalisation is conducted through the seeming sobriety of self-surveillance, rather than through older public spectacles of physical brutalities, such as those observed in the slavery punishment system. This more modern discipline is regulated by those other than the subject (e.g. the nurse, the angry neighbour, the eavesdropper, the pastor, the priest), but first and foremost by the subject themselves through self-surveillance. The enforcement of disciplines occurs not only through the mechanisms of health care interactions, treatment or education, but also through the "micro-physics" of interactions in the multiple spaces of the everyday. For Foucault, these micro-physics function as a network supported by and working as an extension of a much larger system of disciplinary mechanisms which widely connect all the major institutions of modernity (Foucault 1977). Such institutions of modernity include, among others, the supermarket, the outside vendor's market and the workplace; the hetero-normative family and the state; the prison and the medical clinic. Through discipline, the occurrence of *allowable deaths* is increasingly subtle as it becomes part of a disciplinary regime of uninterrupted governmentalities of valuation in relation to human life.

Conclusion

One underlying conclusion drawn from the study described here is that from a human rights perspective the right to life reminds us that there is an inherent value in all life. The right to health, education, housing and food then further infers that there are positive steps that government should be prepared to take in ensuring the protection of the right to life of all people, including people infected and living with HIV/AIDS. However, as the study shows, if the enjoyment of the rights to life and health for people is conditional upon and based on wealth, the ability to pay or CBA calculations, then human lives are not of equal value. It also follows that poorer people living with HIV/AIDS, lesbian, gay, bi-sexual and transgender (LGBT) populations, those living in rural areas and prisoners living with HIV are much less protected by rights to life and health. These groups suffer discrimination, abjection and disadvantage and experience a higher risk of *allowable deaths*.

Furthermore, while the valuation of human life is not an explicit process in health policy, rationing decisions of available HIV/AIDS resources using CEA or CBA calculations, among others, implicitly value and valorise human life. By their nature, enforcing rights to life and health involves rationing of finite resources. Senior government officials in the MoHCC who decide on interventions are implicated here in the production of proxy indicators which correspond to the value of different groups in the population. But equally caught up in this valuation of life are donors who decide which interventions will be funded in each 'envelope' of aid. The Ministry of Finance (MoF) also has a critical role in allocating national aid. While health ministry rations are an already determined share of the state budget, the MoF makes decisions in relation to who gets what among sector departments. Simply, it can decide who deserves more attention and who does not. The key issue in allowing death is not whether the right to life (and the inherent value of human life) is applicable, because it will always be applicable, but rather, what choices decision-makers and health care providers including in the MoF make to preserve the lives of poorer social groups, such as those living with HIV/AIDS.

The cases of rare and fatal diseases, e.g. the hopeless coma, the vascular events that steal brain function, remain difficult to confront. However, society's first response should be to mirror our best instincts as humans, incorporating memories and experiences to care for those in dark times, and not to give the message (by use of mathematical and economic calculations) that those who feel fragile and useless are right to think so and their worth/value is not equal to ours. In this sense, and in contra-distinction to what modern economic calculative devices may claim, human life has a value greater than choices from calculative rationalities, not least because these are themselves shaped by the logics of capitalism. Charity and Tino, however, demonstrate here that the value of human life is intimately connected to social and cultural norms and values. As a humanist, it is possible to note that the manner in which society has treated Charity and her family has conferred on them a low worth, despite the principle that our value should always exist even when we cease to see it.

References

Alberti R. and Emmons M. (2017) *Your perfect right: assertiveness and equality in your life and relationships* New Harbinger Publications, Oakland, CA.

Alston P. (ed.) (1999) *Promoting human rights through bills of rights: comparative perspectives* Oxford University Press, Oxford.

Amar A.R. (1998) *The Bill of Rights: creation and reconstruction* Yale University Press, New Haven, CT.

Belshaw C. (1997) "Abortion, value and the sanctity of life" *Bioethics* 11: 130–150.

Bhat N., Kilmarx P.H., Dube F., Manenji A., Dube M. and Magure T. (2016) "Zimbabwe's national AIDS levy: a case study" *SAHARA-J: Journal of Social Aspects of HIV/AIDS* 13: 1–7.

Bynum W. (2008) *The history of medicine: a very short introduction* (Vol. 191) Oxford University Press, Oxford.

Campbell C., Foulis C.A., Maimane S. and Sibiya Z. (2005) "'I have an evil child at my house': stigma and HIV/AIDS management in a South African community" *American Journal of Public Health* 95: 808–815.

Castro A. and Farmer P. (2005) "Understanding and addressing AIDS-related stigma: from anthropological theory to clinical practice in Haiti" *American Journal of Public Health* 95: 53–59.

CDC(CentersforDiseaseControlandPrevention)(2011) *USCentersfor DiseaseControl and Prevention* Vaccine Injury Compensation Program [Online] https://www.cdc.gov/vaccines/pubs/pinkbook/downlads/appendices/d/vicp.pdf (Accessed 4 July 2017).

Coleman C.L. (2017) "Health related and depressive symptoms among seropositive African Americans" *Applied Nursing Research* 33: 138–141.

Coleman C.L. and Holzemer W.L. (1999) "Spirituality, psychological well-being, and HIV symptoms for African Americans living with HIV disease" *Journal of the Association of Nurses in AIDS Care* 10: 42–50.

Crawford R. (1994) "The boundaries of the self and the unhealthy other: reflections on health, culture and AIDS" *Social Science and Medicine* 38: 1347–1365.

Culyer A.J. (1989) "The normative economics of health care finance and provision" *Oxford Review of Economic Policy* 5: 34–58.

Dolan P. (2008) "Developing methods that really do value the 'Q' in the QALY" *Health Economics and Law* 3: 69–77.

Dworkin R. (1978) *Taking rights seriously* (Vol. 136) Harvard University Press, Boston, MA.

Edwards D.M., Shachter R.D. and Owens D.K. (1998) "A dynamic HIV-transmission model for evaluating the costs and benefits of vaccine programs" *Interfaces* 28: 144–166.

Eibach U. (1998) *Sterbehilfe Tötung aus Mitleid? Euthanasie und "lebensunwertes" Leben* R. Brockhaus Verlag GmbH Brockhaus, Munich.

Feldman D. (2002) *Civil liberties and human rights in England and Wales* Oxford University Press, Oxford.

Foucault M. (1977) *Discipline and punish* Pantheon Books, New York.

Foucault M. (1980) *Power/knowledge: selected interviews and other writings 1972–1977* Pantheon Books, New York.

Foucault M. (1982) "The subject and power" *Critical Inquiry* 8: 777–795.

Friedman D. (1982) "What is 'fair compensation' for death or injury?" *International Review of Law and Economics* 2: 81–93.

Goudge J., Ngoma B., Manderson L. and Schneider H. (2009) "Stigma, identity and resistance among people living with HIV in South Africa" *SAHARA-J: Journal of Social Aspects of HIV/AIDS* 6: 94–104.

Holmes D. (2013) "Report triggers quibbles over QALYs, a staple of health metrics" *Nature Medicine* 19: 248.

Hove-Musekwa S.D., Nyabadza F., Mambili-Mamboundou H., Chiyaka C. and Mukandavire Z. (2014) "Cost-effectiveness analysis of hospitalization and home-based care strategies for people living with HIV/AIDS: the case of Zimbabwe" *International Scholarly Research Notices* 1–13.

Hughes D. and Griffiths L. (1997) "'Ruling in' and 'ruling out': two approaches to the micro-rationing of health care" *Social Science and Medicine* 44: 589–599.

Klein R., Day P. and Redmayne S. (1996) *Managing scarcity* Open University Press, Buckingham.

Machingura F. (2016) *Allowable death and the valuation of human life: a study of people living with HIV and AIDS in Zimbabwe* Unpublished PhD thesis, Global Development Institute, University of Manchester.

Mishan E.J and Quah E. (2007) *Cost–benefit analysis* Routledge, New York.

MoHCC (Ministry of Health and Child Care) (2016) *The Zimbabwe population-based HIV impact assessment 2015–2016* Ministry of Health and Child Care, Harare.

OED (2003) *Oxford English Dictionary*, Oxford.

Parker R. and Aggleton P. (2003) "HIV and AIDS-related stigma and discrimination: a conceptual framework and implications for action" *Social Science and Medicine* 57: 13–24.

Porter R. (1999) *The greatest benefit to mankind: a medical history of humanity* WW Norton, New York.

Rothman D.J., Marcus S. and Kiceluk S.A. (eds) (1995) *Medicine and Western civilization* Rutgers University Press, New Brunswick, NJ.

Sacks D.W., Stevenson B. and Wolfers J. (2012) "The new stylized facts about income and subjective well-being" *Emotion* 12: 1181.

Scheunemann L.P. and White D.B. (2011) "The ethics and reality of rationing in medicine" *CHEST Journal* 140: 1625–1632.

Slovic P. (2010) "'If I look at the mass I will never act': psychic numbing and genocide" in Sabine Roeser *Emotions and risky technologies* Springer, Dordrecht, the Netherlands, 37–59.

Ssenyonjo M. (2008) "State reservations to the ICESCR: a critique of selected reservations" *Netherlands Quarterly of Human Rights* 26: 315–358.

UN (United Nations) (1945) *United Nations Charter: preamble, purposes and principles* United Nations, Geneva.

UN (United Nations) (1948) *The Universal Declaration of Human Rights* United Nations, Geneva.

UNHR (United Nations Human Rights) (1966a) *International Covenant on Civil and Political Rights* United Nations, Geneva.

UNHR (United Nations Human Rights) (1966b) *International Covenant on Economic, Social and Cultural Rights* United Nations, Geneva.

ZIMSTAT (Zimbabwe National Statistics Agency) and ICF International (2015) *Zimbabwe Demographic and Health Survey 2015: Key Indicators* Zimbabwe National Statistics Agency and ICF International, Calverton, Maryland, Harare.

5 Valuing infrastructure

Competing financial and social valuations in the South Durban port expansion

Sarah Bracking and Aurora Fredriksen

Introduction

For several years now, a major expansion of South Africa's port in Durban has been planned. Proposed by Transnet, eThekwini Municipality and the Government of the Republic of South Africa and using the land of neighbouring Indian farmers, this multibillion-rand infrastructural development project would entail a major artificial 'dig-out' port on the site of the current disused airport. It also involves expansion and deepening in the existing bay-head, the building of new wharfs, the construction of a new oil and petrochemical transport and storage facility, possible expansion of the two existing refineries and linking to a new major transport corridor and pipeline to Johannesburg. The plans entail substantial environmental disruption, including the destruction of an estuarine sandbank of particular importance to spawning fish, frogs and birds, and necessitating the demolition and relocation of the residents of surrounding neighbourhoods, including the historic Clairwood and parts of Wentworth and Umbilo. While for the powerful actors supporting the port expansion the project's value is expressed in terms of development, economic growth and job creation, as well as international competitiveness in a global economy, the plight of those who will be displaced is seen as unfortunate at best and forgettable at worse. This is a 'keyhole-type' case study to explore how financial and social valuation practices compete in relation to an infrastructure project.

The African Development Bank, through its Programme for Infrastructure in Africa brochure, summarises the general 'case for' infrastructure in Africa as follows:

> Closing the infrastructure deficit is vital for Africa's economic prosperity and sustainable development. Improved infrastructure would facilitate increased intraregional and international trade, reduce the cost of doing business and enhance Africa's competitiveness within itself and in the global economy as well as act as a catalyst to Africa's economic transformation and diversification through industrialisation and value addition and sustainable and inclusive growth.
>
> (AfDB 2011)

Major speeches in support of the port expansion, including by the then Minister of Finance in Clairwood in 2012, borrow from all these signifiers and axioms. All the key tropes of neoliberal economics – growth, competitiveness, inclusiveness, sustainable development and so forth – are on display in the case for the value of the 'back of port' (BOP) expansion, as the collection of separate but related projects became known, expressed by these political leaders and mainstream economists. Meanwhile, a loose environmental opposition, most active between 2012 and 2016 (when the plans were postponed), featured the South Durban Community Environmental Alliance (SDCEA), groundWork (Friends of the Earth South Africa) and BirdLife SA as central nodes. Opposition to the BOP expansion counterposed the promotion of the values of inclusive, people-centred development, ecological justice and environmental sustainability. Instead of shaping into a neat binary with big capital and government facing off against environmentalists and residents, however, the articulations of value in this case were multiple, often contested, fragmented, temporary and variously substantiated. The promise of employment, a much-vaunted element of the BOP plans and regularly lauded by ANC-affiliated politicians, had a beguiling effect in an otherwise economically disadvantaged community, despite the widespread destruction of homes that would accompany it.

Theorising infrastructure

In each of the case studies funded under the "Human, non-human and environmental value systems: an impossible frontier?" research programme, material infrastructures play a role in the valuation practices and performative values under consideration. This is not surprising, of course, if one takes systems of accounting through which economic valuations are achieved as infrastructural forms (e.g. Mennicken and Sjögren 2015). However, even leaving aside these accounting infrastructures, our various case studies also involve infrastructures in the more traditional sense of the term. These range from large, fixed assets such as ports and irrigation systems, to the more dispersed material artefacts of modern governance associated with international aid bureaucracies and national healthcare systems, and the material networks of cables and servers that enable the hyperflows of contemporary finance. Moreover, despite the wide range of infrastructural forms involved in our various cases, in all of these, infrastructure is not merely incidental – one of many socio-material actants in an assemblage – but plays a distinct role in the valuation processes.

While space restrictions do not allow a full examination of the relationships between value and development across the relevant cases in this volume, in this chapter we use Durban's proposed BOP expansion as an exemplar to explore the heterogeneous relationships between value and infrastructure in the development field. The chapter begins by situating Durban's BOP expansion within the larger institutional assemblages and narratives through which new infrastructure is framed as essential for development in Africa generally and South Africa in particular. To demonstrate how infrastructures shape the possibilities and realities of value, we explore three related avenues here: first, the durable materiality of infrastructural forms; second, the affectual attachments that infrastructure can inspire; and third, the feedback loops within structures of power and political economy that link powerful political and economic actors with powerful infrastructures. A concluding section reflects on the emergent *dispositif* of infrastructure and value in Durban.

Development value, Africa's 'infrastructure gap' and Durban's proposed port expansion

Since 2008, the issue of the infrastructure gap has come to the fore in Africa, in discourse if not in direct material form. In July 2012, just as the Rio+20 Earth Summit for Sustainable Development was hitting the headlines from Brazil, the huge *Infrastructure Africa* conference was also taking place in Johannesburg. Somewhat ironically, the first was focused on debating the curbing of carbon emissions to avert further global warming, while the second, using much larger proposed budgets and personnel, was in the process of committing African governments to decades more carbon emitting fossil fuel-based infrastructure for cars and coal mines, aviation expansion and container shipping. Indeed, in Africa it appears that the only thing more 'much-needed' than foreign direct investment (FDI) is infrastructure, with the World Bank promoting the view that there is a financing 'gap' worth a wide range of apparently vast sums of money – cited as USD 93 billion per annum in 2012 in the run-up to the Infrastructure Conference (Nepad Business Foundation 2012). The African Development Bank and African Union Programme for Infrastructure Development in Africa (PIDA) estimated (in a typically titled press release "Closing the Infrastructure Gap vital for Africa's Transformation") a 'much-needed' spend of USD 360 billion to close the 'gap' between 2011 and 2040. Realising this spend would in turn require extensive public private partnerships (AfDB 2011). The PIDA Priority Action Plan (PAP) of 51 'infrastructure back-bone' projects alone require USD 68 billion by 2020, over USD 40 billion (or 60 per cent) for energy and over USD 25 billion (37 per cent) for transport (AfDB 2011). Meanwhile, the super new Asian Infrastructure Investment Bank launched as an initiative from China with 37 country members to date, had its first Board meeting in 2016 and has been dubbed the new 'World Bank' for Asia.

Ouma noted that since the financial crash of 2008, financiers have shown a "new love affair with real things" (2014, 162). In this case, the development banks' love affair with infrastructure is shared by the Government of the Republic of South Africa, which houses Special Infrastructure Projects (SIPs) in the Office of the President. President Jacob Zuma even proposed a 'fast-track' for the environmental assessments of SIPs, which are now managed from the executive arm of government in the Presidential Infrastructure

Coordinating Commission, following the passing of the controversial Infrastructure Development Act into law in 2014 (Government of the Republic of South Africa 2014). There were 18 SIPs identified in the 2012 Infrastructure Plan and the BOP expansion is the second most important by value, only topped by a massive coal railway and port development at Richard's Bay. It is South Africa's biggest single location investment project worth upwards of an initial estimate of USD25 billion. Plans to expand Durban's port capacity from around 2–3 million to 20 million twenty-foot equivalent units (TEUs, a unit of cargo capacity used in shipping) annually, include: a port deepening, to accommodate the new super post-Panamax ships of 15,000 containers; a new artificial 'dug-out' port on the old airport site, which will also be used by the petrochemical industry and three back of port refineries; a new offshore oil buoy; an expanded pipeline to Johannesburg (largely completed and doubled in capacity); and extensive back of port logistics and freight facilities requiring the forced removal of thousands of families from the historically disadvantaged communities of Clairwood, Merebank, Wentworth, Isipingo and Umbilo.

Notably then, in the infrastructure gap conception promoted by the international financial institutions and the political elite of South Africa (and elsewhere), the infrastructure in question is generally large, project- rather than programme-based and directly related to mining and industrial capital, rather than to the infrastructure associated with utilities and peoples' residential spaces. Thus, the development in this is also not directly concerned with meeting immediate needs (e.g. for clean water, household energy or sanitation), so much as with supplying the assumed and abstracted human benefits of minerals extraction and industry, found in employment creation and 'trickle down' wealth effects from economic growth. We now explore this value of infrastructure in development through its materiality, its affects and the feedback loops of power.

The durable materiality of infrastructure

Materiality is a definitional feature of infrastructures. As Larkin writes "Infrastructures are matter that enable the movement of other matter" (2013, 329): pipes and irrigation channels that carry water; hospitals, equipment and a trained medical workforce through which healthcare is delivered; networks of cables and servers that enable financial flows; ports that facilitate commercial shipping; and so on. In this respect, the material forms of infrastructures embody very particular use values: they enable movement and circulation. Or, at least, they hold the *promise* of such a use value and all the attendant values that might be generated through such a usage. Not all infrastructures function as intended all of the time; when the material presence of infrastructure falls into disuse or disrepair it may signal to observers the simultaneous promise of value and the failure of this promise (Khan 2006). This paradoxical condition is especially apparent in development, where building new infrastructure is widely assumed, often a priori, to bring value to developing nations despite the dubious longevity and widely unequal distribution of its costs and benefits in many cases (Bracking 2016; Desai 2015). Just as Loftus and March (2015) noted, however, it is the imprint on the ground that is perhaps most significant in this type of large project, which seems to thrive in conditions of financialisation (Bracking 2016). Empty stadiums and empty ports may not materialise development value for most of the population, but in feeding investors' pockets from public taxes, they materialise financial value. Thus, the "very real ways in which a material imprint from financialisation has been left in the landscape is important" (Loftus and March 2015, 175).

Indeed, infrastructural builds with protected derivative income streams are both a dream come true to financiers, and a means of growing finance. Torrance's (2009) work on a new Thames Water desalination plant, for example, showed the reordering of values such that in building the new plant, providing finance to investors was prioritised over providing water to customers. Apparently, this plant may not produce water and may only be used in a drought, but holds inflation-protected returns for institutional investors. Rather than water supply, a "more profound motivation seems to be the need for new infrastructural forms within which to ensure speculative gains" (Loftus and March 2015, 175, summarising Torrance 2009). Hildyard (2016) would support the view that infrastructure, and in particular public–private partnership models, have an overriding aim to create a contractual derivative income stream for investors. In the BOP case certainly, the expected derivative income streams from the build itself in procurement contracts, as well as those expected by the lenders to the state-owned enterprise (SOE) Transnet became early points of contention, just as much, if not more so, than details of the bricks-and-mortar effects on the landscape and people.

In this debate on derivative benefits, economic calculations invariably employ calculative multipliers abstracted from models of growth effects seen elsewhere or generically. This was so in the economic impact assessment of the BOP, as we see below. Indeed part of the allure of infrastructure as a tool for development lies in the assumption of scalability – that is, its ability to be transplanted, expanded or replicated without transformation of the elements (or relationships between elements) involved (Tsing 2012). Because the material forms of many infrastructures are apparently scalable, the relationships of value that these forms have engendered in some places are also assumed to be scalable as they are replicated elsewhere. Larkin captures this sentiment, writing:

> Many infrastructural projects are copies, funded and constructed so that cities or nations can take part in a contemporaneous modernity by repeating infrastructural projects from elsewhere to participate in a common visual and conceptual paradigm of what it means to be modern.
>
> (Larkin 2013, 333; see also Mrázek 2002)

Achieving this type of scalability, of course, involves denying, or 'framing out' in our more specialised conceptual taxonomy, the differences and heterogeneity between places – the *non*scalable elements and relationships (Tsing 2012) – that prevent identical infrastructural projects from achieving identical development value across different times and places. The Durban BOP expansion was certainly officially described, evaluated and marketed as a scalable development, replete with mentions of global, international and shipping 'best practice' and many references to large-scale, efficiencies of size, modernist drawings of neat docks and, of particular political importance, over 27,000 new jobs.

Alongside the promise (and potential failure) of scalable developmental and economic values, the materiality of infrastructure can embody certain relations between governments and populations. Deville and colleagues (2014), for example, lay out the ways in which national governments' conceptualisations of risk and preparedness are enacted in the durable structures of emergency shelters. They consider how the "particularly durable, supposedly permanent aspect of concrete" used to construct shelters acts as a potent and enduring material symbol and embodiment of the relationship between states and populations, a relationship they call 'concrete governmentality' (Deville et al. 2014, 186). As seen in Chapter 2, the term governmentality indicates a political rationality shaping the 'conditions of possibility' (Collier 2009; Foucault 2003) for change at any particular moment. Here, adding concrete is both a play on the ubiquitous construction material and its pervasive role in changing landscapes and the built environment, as well as being a metaphor for and an indication of the way built infrastructures solidify certain relationships between governments and populations (see also Harvey 2010). An important subset of this concretisation of relations is that which reifies the uneven relationships between governments and different populations. This is the case, for example, when water infrastructures reliably pipe water to some users and not to others (Peters and Woodhouse, Chapter 11 this volume; Anand 2011, 2012; Von Schnitzler 2008); or when healthcare infrastructures act to systematically deny certain populations adequate healthcare (Machingura, Chapter 4 this volume). Certainly in the peripheral townships and informal settlements of eThekwini (the metropolitan area in which Durban is the anglophone-named city), a concrete road, drain, sewer pipe or tap are widely perceived as the indicators of entry into full citizenship within the metropole.

Instead of this concretely symbolic inclusion, however, the operational plans for the Durban port expansion will require the removal of a large number (currently contested but likely in the tens of thousands) of settled families in Clairwood, Isipingo, Merebank and some in Umbilo and Umlazi, traditionally black, 'coloured' and Indian neighbourhoods. These communities came to rest here following exit from colonialist indentured labour regimes and prior apartheid forced removals. Thus, the impact of plans for the BOP expansion reflect the path dependence of prior histories of racial and economic inequality as well as the infrastructures already in place. The expansion plans for the freight corridor, new logistics centres, new dug-out port and increases in road traffic impact most on communities already living with dirty industry and an ad hoc experience of traffic expansion and informal industrial sprawl. Under apartheid, dirty industry, landfill dumps, toxic polluters and the like were placed proximately to disadvantaged communities of colour. Rather than addressing these material and spatial inequalities, the BOP expansion further concretises already unequal relations; building on, taking advantage of and expanding this discrimination. The rerouting of the oil pipeline from the Durban oil refineries to Johannesburg, already completed, attests to the durability of racialised environmental injustice, as its original plans were changed to incorporate more travel through black, as opposed to white, areas.

However, the rewriting of old relationships is not the only issue in evidence here. A re-recruiting of an assemblage of constituencies, loose alliances, interest groups and identifiers is also at play, communicated through the literature and documentation of the BOP expansion plans and policy engagements. Therein are notable silences of relationships, evasions and denials that relationships exist at all, and material abjections of prior relationships, in the sense described by Ferguson (2006) of a throwing down or throwing out of persons. These erasures are performed through projected visions of how the future will look if the BOP expansion is completed. In other words, if the expansion goes ahead, the communities of Clairwood, Merebank and Isipingo will be largely dissolved. In anticipation of this, or in a performativity which suits the outcome, these communities are largely absent from the planning documents. For example, the impact assessments drawn up by consultants read like promotional literature in support of Transnet and its project. In fact, the key impact assessment already 'zones' out the people and the names of the communities affected in the title: *A Local Area Plan and Land Use Management Scheme for the Back of Port Interface Zone* (Graham Muller Associates 2009a). Rather than specific communities whose material presence will be erased and residents forcibly displaced, these residents have become here merely an 'interface zone' for the material expansion of the port.

This material embodiment of government relations with a population is closely tied to the second way in which we suggest infrastructures shape values and valuation practices: the affective responses they elicit.

Affect and infrastructure

The materiality of infrastructure has often been framed as largely invisible, only becoming visible when it breaks down (e.g. Star 1999; Collier 2011). This can be illustrated in the networks of cables and warehouses of servers that enable finance. However, the materiality of some infrastructures – including most large-scale development projects like roads, dams, ports and so on – is not only highly visible, but the material visibility of these projects is an essential aspect of what Larkin (2013) calls their 'political effects'. This is the case where the material visibility of infrastructures is played as spectacle, deployed to elicit powerful affects and emotional attachments.

Harvey and Knox (2012) suggest that enchantment, in the sense meant by Bennett (2010), explains how large-scale, highly visible infrastructures can maintain a 'generic social promise', even when such a promise can be seen to have failed in certain aspects. Larkin similarly notes the 'fetish-like' aspects of some infrastructures that elicit 'forms of desire and fantasy', wholly separate from the technical functioning (or malfunctioning) of the said infrastructure (2013, 329). Many of these affectual attachments to large-scale infrastructures are intended or 'inscripted' by those designing or commissioning them (Akrich 1992; Marres 2012). Todorov (1994) has shown, for example, how Soviet factories technically underperformed, yet still achieved the symbolic meanings intended by communist governments (see also Mrázek 2002). In the field of development, new infrastructures are routinely inscripted with ideas of progress and modernity. Linking back to the use-value of infrastructures as material forms that enable the movement or circulation of things, Larkin summarises how this promise of use-value engenders a promise of other, less material values:

> by promoting circulation, infrastructures [are imagined to] bring about change, and through change they enact progress, and through progress we gain freedom. Perhaps this process explains why as objects they provoke such deep affectual commitments, particularly, but not only, in developing societies.
>
> (Larkin 2013, 332)

Large and highly visible infrastructural projects often capture the imagination of publics who see them as valuable even without a clear concept of the development value (or lack thereof) in these projects (see De Boeck 2011 for a striking example). In this sense, the lack of a clear economic projection of value is not necessary and does not matter. This is because infrastructures are not simply attached to economic values – though these are often assumed – but are central to political imaginaries of progress and modernity (Harvey and Knox 2012; Larkin 2013). As Desai (2015) and Dyer (2014) summarise, the economics of investing ZAR 250 billion in Durban, a container port where costs are already the highest in the world, do not make standard economic sense, but:

> Once more, like King Shaka airport, Moses Mabhida Stadium, and the International Convention Centre, ratepayers will have to make up the shortfall for massively underutilized infrastructure whose chief benefit is to those who get the contracts to build them and the political class who serve as the "business partners".
>
> (Desai 2015, 25)

Desai terms this 'Faustian development' (ibid. 2015, 27). But, despite the lack of clear development value for most of the population, these infrastructures elicit powerful affects through their situation in the ruling discourse of patriotic, or populist, nationalism, such that to oppose these iconic projects is to "go against the national interest" (ibid. 2015, 32).

The example of the FIFA World Cup 2010 infrastructural funds is an infrastructure built for affect in the service of nationhood and identity, as opposed to sound economic or development value. These funds built new stadiums in South Africa in a model that transferred risk and cost to future taxpayers, while short-run income streams were guaranteed for 'tenderpreneurs' and offshore financiers. As Desai writes of the Moses Mabhida Stadium built in Durban for the 2010 FIFA World Cup:

> Global finance capital ... thrives in a world of large-scale investment in mega-projects, mega-events, and the short-term investment, long-term debt, and creative financing associated with them. The model also imposes a world view in which the West becomes the unit of global measurement.
>
> (Desai 2016, 90)

The Moses Mabhida Stadium, where local football teams regularly draw crowds of just a few thousand, seats 70,000 people. Meanwhile its standard operating costs, excluding renovations and repair, are ZAR 88.5 million a year, with only ZAR 9 million retrieved from 'non-bowl' activities such as shops, restaurants and a gym. Thus, the stadium must make over ZAR 1.5 million per week to break even (the remaining ZAR 79.5 million), and yet "[a]s anyone can attest who regularly passes by the great empty hulk of a building ... this level of use is nowhere near being attained" (Desai 2016). Thus the whole infrastructure exercise, underpinned by cost–benefit calculations privileging short-term derivative income flows to top tranche investors and the 'tenderpreneurs' over long-run costs to the taxpayer, was made popular in Durban through a language of 'mega-projects', the 'Madiba Magic' of global inclusion and modernity (Desai 2015, 2016).

In the BOP expansion case, and in others like it, the appeal to modernity and affect is entangled with discourses of economic and development value, with these values acting through a "heavy infrastructure of conventional forms ... formal indicators, accounts and procedures, standards, benchmarks, good practices, etc." (Thévenot 2015, 195). It is a language of formalisation (ibid.), which was used by many parties to the debate and contestation over the port development in 'critical moments' where an appeal to justification was required (Boltanski and Thévenot 1999, 359). These standpoints are situated in a messy configuration, which is best described as a *dispositif* – an emerging or contested arrangement of correlations between things, people, non-human species and discursive and calculative devices in place (see Braun 2014, 50–51; citing Foucault 1980, 194). This arrangement is different to the more traditional understanding of struggles around infrastructure, where a configuration of 'good environmentalists-versus-big capital' is depicted, with two standpoints which are mutually exclusive and necessarily competing (more below).

In reflecting on the performance of nationhood and patriotism through South Africa's mega-project infrastructures, Desai quotes Castells's point that "the flows of power generate the power of flows, whose material reality imposes itself as a natural phenomenon that cannot be controlled or predicted … people live in places, power rules through flows" (Desai 2016; citing Castells 1989, 349). This brings us to the third way in which infrastructures are implicated in the articulation and performativity of values: namely, the feedback loops of power.

Infrastructure and the feedback loops of power

Çalışkan and Callon write that market inequalities result from "the unequal power of calculating agencies that loop back to reinforce themselves. Due to these asymmetries, the most powerful agencies are able to impose their valuations on others and consequently to impact strongly on the distribution of value" (2010, 13). This general assessment of market values is useful for thinking about power, value and infrastructures. As in markets, asymmetries in power allow some infrastructural assemblages to self-reinforce their own value at the expense of other infrastructural forms. In the material scalability of infrastructures and the grand political theatre of mega-projects affectually framed as development, other, smaller, infrastructures are inevitably framed out: infrastructures that are local, pre- or a-modern infrastructures of the powerless. In Durban's Clairwood neighbourhood, various informal social, economic and cultural infrastructures stand to be swept away by the BOP expansion and the value of this existing infrastructure denied. Indeed, the local government structure and the planning documents it produces do not see the extent of the informal economy as they are spatially and culturally removed from the poor (Tshabalala et al. 2017).

Of course the fact that inequalities in markets produce asymmetries of *power between and within* market agencies and institutions is related to the '*outside*' of market structures and what happens there in institutions of government and society. In the BOP case, market inequalities are themselves authored by the mode of regulation adopted by the political elite and President Zuma and his 'state capture' cronies in particular. As has now been described eloquently by the State Capture Research Project (2017), during 2010 to 2016, President Zuma and a senior group of connected persons within government and in the private sector, particularly the Gupta family, successfully promoted a political project under the rubric of radical economic transformation, which repurposed government and redirected the procurement spending of the SOEs to themselves. The spending begun and anticipated for the BOP, like other SIPs, was placed under the Presidential Infrastructure Committee and was to be largely spent through Transnet.

This committee and other inter-ministerial committees secured 'parallel government and decision-making structures' (2017, 20), which, according to the authors of the State Capture Research Project, was important in centralising power under President Zuma and which (among other developments) contributed to making state capture possible (2017, 19–20). Committees such as these lack transparency and do not have to report to Parliament, as individual cabinet members must, and thus constitute a thinning of parliamentary oversight (Democratic Alliance 2014). While the details of the politics are not needed here, the overall pattern is important, since the structures of power under a situation of state capture by a power elite are synergistic to the promotion of big infrastructure projects. Such projects are ideal conduits for the flows of spoils.

Apart from the opportunities afforded for personal enrichment, there are several other reasons for the popularity of big infrastructure spending from the perspective of the political economy of elites. First, in the absence of an industrial policy that is working to produce jobs, building something that creates jobs at opportune moments in the fiscal and electoral cycle has obvious benefits for those seeking re-election. Second, in a country with racialised inequality, construction, where start-up costs are relatively low, is one of the few economic sectors where historically disadvantaged persons can gain entry. Consequently, jobs can be distributed among a constituency within an 'economic justice' discourse, and to key ruling party supporters in exchange for political finance. Third, public procurement and tendering processes can be further designed to assist new market entries, political support for the ruling party and to raise political finance. Fourth, justifying public expenditures on infrastructure can be done principally by referencing the beneficiaries of procurement, building and employment created in these processes, while the actual long-run costs in loan repayment and sustainable jobs which remain after the construction (often much fewer) are downplayed. Thus in terms of financialisation processes overall, there is also a synergistic relationship between building infrastructure; concentrating power within party-states and the greater extractivism made possible for financiers and construction companies from the national economy.

Unsurprisingly, the consultants (Graham Muller Associates Consortium and UrbanEcon), who were commissioned by eThekzwini Municipality to conduct impact assessments, assumed the BOP expansion was in the national interest a priori, such that adverse effects on persons and nature were assumed to be unavoidable and necessary, if they were mentioned at all. UrbanEcon used textbook growth multipliers, normal when scalability is assumed, to argue for surprisingly precise job creation figures of "318 279 temporary jobs created through the port construction process … 63 656 jobs during a 30-year construction phase … job creation of an additional 27 533 jobs at full operation" (Feigenbaum 2011). Meanwhile, to measure not jobs, but the size of formal business' contribution to GDP, Graham Muller Associates conducted a 'status quo assessment' of the back of port area, thus comparing a future scenario with a status quo that will no longer exist, and leaving unmeasured the future of the proximate people affected (2009b).

With the power of international development banks, the South African government and its hired consultants all feeding back into these narratives of immense value creation through the BOP expansion, there has been little room in official accounts for opposing values to be voiced. This is despite the fact that the economics used to support the BOP expansion failed to convincingly make the case for its economic value on a number of counts (Bracking 2013a, 2013b). First, a comparative sector analysis was not carried out. Top-down planning suggests that all is possible (win–win), but the interests of different economic sectors can contradict each other, with, in this case, possible conflicts with leisure, tourism, cultural industries, sports and lifestyle industries, retail and small commercial, and light industrial ventures. Also missing was a comparative counterfactual analysis that would look at the opportunity costs of investing in alternative job-creating projects – as Desai points out, even if the jobs promised in the consultant reports were created they would cost ZAR 1.9 million each (2015, 26). Most critically, perhaps, other increases in regional capacity were not considered. These include: the Bolloré Africa and CFM (ports and railway parastatal) building Mozambique's first oil port in the city of Pemba; Grindrod, Africa's largest shipping company, and Mozambique's Maputo Port Development investing USD 1.7 billion over the next five years to upgrade ports; the tripling of capacity at the Maputo and Matola ports from 15 to 50 million tons by 2020; port expansion in Angola at Lobito (USD 1.2 billion investment by government), Cabinda and Luanda (over USD 350 million); a multimillion-dollar greenfield port at Lamu in Kenya; and the planned USD 11 billion Bagomoyo port in Tanzania which will be bigger than the Dar es Salaam and Mombasa ports, both of which are also expanding.

In addition, the externalities of the costs of the dig-out port were not costed into the BOP expansion's official accounts of value, including costs to people (social and economic impact, heritage capital) and the environment. These were significant, and only inadequately anticipated. For example, a minute of the eThekwini Housing Committee in May 2012 noted that 12 settlements, involving 30,000 individual units excluding the 'backyard shacks', would need to be relocated into the housing development at Cornubia. At that time this was in its infancy and would include only 12,000 units when complete, such that there was "an oversubscription of almost 18 000 units" (22 May 2012, Head of Housing to ECOD Housing Committee). In addition to this social cost, Graham Muller Associates (2009a) also noted environmental effects:

> A significant area will be displaced, compensation may require remaining areas of coastal grassland such as the racecourse in addition to significant areas outside the area. The loss of habitat associated with port development may not be replaceable in the location.

Initially, 'conserving' the Clairwood racecourse was proffered as an 'offset' for this cost, but by August 2015, the racecourse had also controversially been transferred to developers for a logistics park, leaving little prospect for offsetting the costs of the BOP expansion borne by the people and environment in South Durban (SDCEA 2014; Hallowes 2014).

In an alternative assessment of income, assets and actual employment already in the area, Diga and Bracking found nearly 500 jobs (of which 491 were mostly informal) existing in just 1,000 representative households surveyed (Diga and Bracking 2015, 16). Were the BOP expansion to go forward, the inevitable job losses were notably not included as a deduction from the projected jobs to be 'created' and widely advertised by the contracted consultants. Moreover, it was discovered that over 11.5 per cent of the residents sampled were in informal (or "illegal") settlements, which were not counted for the resettlement considerations (Diga and Bracking 2015, 9). The alternative assessment further found that 23.4 per cent of households were living on less than ZAR 1,000 per month; 57.3 per cent earned less than ZAR 3,000 (Diga and Bracking 2015, 8); and only 43.4 per cent were receiving social assistance, 87.8 per cent of which was from the government (Diga and Bracking 2015, 18, 19). At the time of the survey, the claimant income level for a Child Support Grant, for two people, was ZAR 2,600 suggesting that a significant portion of people who should have been able to collect assistance were not able to. In the survey, 18.1 per cent were successfully claiming Child Support, and 17.2 per cent an Old Age Pension (2015, 20). Thus the people and their environment have been historically neglected, and are not currently 'seen' by a state elite and their contracted consultants. The poverty of the community was instead used against them to suggest that what would be lost was not worth saving, despite huge cultural, social and historical infrastructure and capital present in the neighbourhood (Bracking et al. 2015).

So, while the case for the economic value of the BOP expansion for the proximate persons was never established, in fact, it can be argued that this calculation was not even attempted. To begin with, the aggregate case was also weak (commensurate with more general findings on the economic projections associated with large projects). More importantly, perhaps, establishing the value of the BOP expansion for those with little power was considered unnecessary from the point of view of the powerful actors supporting its construction. The valuation processes used to justify infrastructure builds do not typically demonstrate a returned profit on a national scale, although they invariably promise jobs and growth. Instead, and reinforcing their existing power, complex global financialisation assemblages have altered financial parameters in costing the benefits of infrastructure in favour of (further) privatisation of profits offshore and socialisation of costs onshore (Bracking 2016). This also illustrates how conventional development economics has difficulty grasping the value of large construction projects in social terms, mostly because it occludes the issue of scale. Assuming a national frame of 'much-needed' FDI in the present, it fails to account for the local and future costs of infrastructure projects.

Dispositifs and unlikely alliances

As of August 2016, the BOP project had been postponed due in part to the global economic slowdown in shipping, but also as a result of problems in financing and resistance by the communities affected (groundWork and SDCEA 2016; Bond 2016). While traditional models of a binary struggle between capital and labour, or polluting industries and environmentalists, may give rise to debate about whether the latter 'won' the struggle, or capital (merely) changed its mind, the case study here could equally reflect what Braun has recently termed a biopolitics of 'eco-cybernetics' (2014, 50), where the administration of life is increasingly integrated and iterative to the global economy and city assemblage. In Braun's understanding, environments and people become 'infrastructures' or the background canvas for a built project, while increasingly large-scale capital assets, such as the so-called 'ecosystems' of co-location warehouses for financial exchanges, are described and conceived in ecological terms. In this reading, the underlying design flaw of the BOP project could have been simply that its overly modernist and imposed style jarred with new framings of city landscapes as resilient and flexible spaces where tourism and industry can be co-located. The heterogeneity and shifting terms of the BOP expansion can then be read as reactive to multiple points of opposition, including from other sections of business. The complexity of variously positioned groups surrounding the BOP expansion may then be usefully seen as a locally situated '*dispositif*' in the Foucauldian tradition, that is, "a thoroughly heterogeneous set consisting of discourses, institutions, architectural forms, regulatory decisions, laws, administrative measures, scientific statements, philosophical, moral, and philanthropic propositions" (Foucault 1980, 194; cited in Braun 2014, 51). Thus, while the scope of the BOP expansion *dispositif* is tilted in favour of big capital by discourses of modernity and structures of power, its modernist design, as demonstrated in the BOP plans, still attracts oppositions as it fails to appreciate the fissures, contradictions and contested valuation practices of groups, agencies and interests in place.

Rather than aligning with the modernist affect, understandings of the value (or lack thereof) in the BOP expansion by those interviewed in Clairwood were multiple and at times at odds. For example, poor residents involved in the trucking industry as drivers, workers, owners of storage facilities or providers of breakdown or salvage services tended to support the BOP expansion plans, despite the vociferous and sometimes near-violent opposition of their erstwhile neighbours (Bracking et al. 2015). For other residents already seeking exit from a polluted Clairwood, where trucks blast past regularly killing bystanders in what is supposed to be a residential neighbourhood, the speculative land market that flourished from 2013–2015 in expectation of the BOP expansion was sufficient to entice them to sell up and relocate of their own accord. Indeed, evidence collected in the Clairwood survey (Diga and Bracking 2015) showed widely fluctuating land and property values in 2014, fed by speculation and uncertainty over present and future values. One interviewee ironically valued his cardboard home at ZAR 3 million, lying as it did in the path of the area to be rebuilt. Another said he wouldn't leave his shack for less than several million rand. Similarly, the SDCEA-led 'we will not move' campaign was somewhat frustrated by the clear views of the extremely poor informal sector and shack dwellers that if the BOP expansion entitled them to assisted removal and rehousing to 'anywhere with a tap' then it would be welcomed (Respondent, Clairwood survey). Others bemoaned that they had already been made violently homeless by landowners clearing informal units to make way for truck parking and land sales. Consequently, an easy coherent coalition either for or against the BOP expansion was not to be found.

To illustrate the multiplicity of values at play, consider two aspects of the Clairwood survey, which were discussed and clarified at a community engagement event at the Sikh Hall in late 2014. First, a very strong resonance of community is found in many of the 1,000 households surveyed, with persons describing a deep sense of belonging in Clairwood (Bracking et al. 2015, 9–10):

> Clairwood is who we are [...] Clairwood is more than just an area, it is our life.
>
> (Male, Sastri Rd)

> Everyone here is in the same struggle here and we share our struggles and that is who we are. I like how there's a mix of all races here: Coloured, Indian, White, Zulu, Xhosa, Zimbabwean, everything.
>
> (Female, in informal settlement)

> Here we would receive help from neighbours even if we do not have food we would never sleep hungry.
>
> (Female, Sir Kurma Rd)

> I was born here, my mother and father died here. I will die here too.
>
> (Female 20130017, Horsham)

Second, however, over 46 per cent of those surveyed said they would be willing to move if an appropriate level of compensation were offered by government, ranging from ZAR 20,000 to ZAR 5 million (Diga and Bracking 2015, 5–6).

Not all factions of 'big capital' supported the expansion either. Notably those who required an operating environment that is environmentally 'clean', such as hoteliers and leisure industry niche operators – in this case Durban Underwater Club, the Yacht Club, uShaka Marineworld and so forth – also have a conflicted relationship to the BOP expansion, given the associated growth of polluting risk and fossil-fuel industrial assets. While necessarily muted by the clear support of the ruling party for the BOP project, these dissenting views were nonetheless in evidence at important 'critical moments' of value justification, such as: in September 2012 when the Finance Minister spoke at the Sikh Hall to an initially compliant audience who could be nonetheless roused into opposition by the polemics of Professor Ashwin Desai; at the Durban Chamber of Commerce discussion chaired by Andrew Layman in July 2013; during the SDCEA discussion at Amanzimtoti Conservancy with Transnet's Marc Descoin in 2013; at the eThekwini EDGE Forum-sponsored discussion event in the Presidential Suite at Moses Mabhida in February 2015, where a maritime expert from the World Bank spoke against the plans and in favour of an incrementalist approach.

Conclusion

In this chapter, we have explored three major avenues through which infrastructure *dispositifs* come to enact value (in multiple forms): materiality, affect and power. Although the values enacted in these ways are sometimes scripted and intentional, they may also have unplanned, emergent effects. Interesting changes to policy and response in the cities of the Global South are occurring, partly because of the infiltration, shared terrain and global scale of development languages and negotiations of value. In eThekwini global knowledge and practice are widespread, while work from eThekwini contributes to the formation of global 'best practice', thus problematising any hierarchical organisation of scale in the knowledge economy and its effectual devices of value justification. In consequence, understanding the values and valuation practices of large-scale infrastructure projects requires an accurate empirical exploration of the *dispositif* of the everyday life of projects, programmes and activities. Thus, while there is a tendency among parts of the traditional 'Left' for moments of struggle to be read in an epiphenomenal way from class location (such as in the proposition that all residents would automatically oppose the build), this is not the case. Nor is it that persons have false consciousness or are otherwise deluded. This speaks only to the arrogance of the observer who knows that their standpoint is invariably correct, the form of critique criticised by feminist theory (e.g. Haraway 1997). Instead, development value must be researched honestly, acknowledging that it is inevitably conflicted by economic injustice, hopes and aspiration as experienced through complex processes of affect, and different views on the nature of progress and change.

References

AfDB (2011) *PIDA, The Programme for Infrastructure Development in Africa: Transforming Africa through Modern Infrastructure.* African Development Bank. URL: www.afdb.org/fileadmin/uploads/afdb/Documents/Generic-Documents/PIDA%20brief%20closing%20gap.pdf (Accessed 21 August 2017).

Akrich M. (1992) "The de-scription of technical objects" in Bijker W. and Law J. (eds) *Shaping Technology, Building Society: Studies in Sociotechnical Change*. MIT Press, Cambridge, MA, 205–224.

Anand N. (2011) "Pressure: the politechnics of water supply in Mumbai" *Cultural Anthropology* 26(4): 542–564.

Anand N. (2012) "Municipal disconnect: on abject water and its urban infrastructures" *Ethnography* 13(4): 487–509.

Bennett J. (2010) *Vibrant Matter: A Political Ecology of Things*. Duke University Press, Durham, NC.

Boltanski L. and Thévenot L. (1999) "The sociology of critical capacity" *European Journal of Social Theory* 2(3): 359–377.

Boltanski L. and Thévenot L. (2006) *On Justification: Economies of Worth*. Princeton University Press, Princeton, NJ.

Bond P. (2016) "Red–green alliance-building against Durban's port-petrochemical complex expansion" in Horowitz L. and Watts M. (eds) *Grassroots Environmental Governance: Community Engagements with Industrial Development*. Routledge, London.

Bracking S. (2013a) "Durban Port's poverty, consumerism, enoughness" *The Africa Report*, 29 April. URL: www.theafricareport.com/Columns/durban-ports-poverty-consumerism-enoughness.html (Accessed 7 August 2017).

Bracking S. (2013b) "Heading for a crash without seatbelts" *The Mercury*, 5 July: 7.

Bracking S. (2016) *Financialisation of Power*. Routledge, London.

Bracking S. and Diga K. with Pia Falschebner, Mandy Lombo, Tawonga Rushambwa and Ayanda Tshabalala (2015) Clairwood Survey: community views of the value of Clairwood in the context of current development plans for Durban port expansion. Research Report 2015 No. 1, University of KwaZulu-Natal and Leverhulme Centre for the Study of Value, https://appliedpovertyreduction.files.wordpress.com/2015/01/emailing-clairwood_qual_report-finished-minus-frequency-table1.pdf (Accessed 21 August 2017).

Braun B. P. (2014) "A new urban dispositif? Governing life in an age of climate change" *Environment and Planning D: Society and Space* 32: 49–64.

Çalışkan K. and Callon M. (2010) "Economization, Part 2: a research programme for the study of markets" *Economy and Society* 39(1): 1–32.

Castells M. (1989) *The Informational City: Information, Technology, Economic Restructuring and the Urban–Regional Process*. Blackwell, Oxford.

Collier S. J. (2009) "Topologies of power: Foucault's analysis of political government beyond 'governmentality'" *Theory, Culture & Society* 26(6): 78–108.

Collier S. J. (2011) *Post-Soviet Social: Neoliberalism, Social Modernity*, Biopolitics. Princeton University Press, Princeton, NJ.

De Boeck F. (2011) "Inhabiting ocular ground: Kinshasa's future in the light of Congo's spectral urban politics" *Cultural Anthropology* 26: 263–286.

Democratic Alliance (2014) *Defending our Democracy: The DA's Plan to Firewall Key Institutions*. DA, Cape Town. URL: https://www.da.org.za/wp-content/uploads/2014/01/Defending-our-Democracy-The-DA%E2%80%99s-Plan-to-Firewall-Key-Institutions.pdf (Accessed 7 August 2017).

Desai A. (2015) "Of Faustian pacts and mega-projects: the politics and economics of the port expansion in the South Basin of Durban, South Africa" *Capitalism Nature Socialism* 26: 18–34.

Desai A. (2016) "Between Madiba Magic and spectacular capitalism: the FIFA World Cup in South Africa" in Gruneau R. and Horne J. (eds) *Mega-Events and Globalization: Capital and Spectacle in a Changing World Order.* Routledge, Abingdon, 81–94.

Deville J., Guggenheim M. and Hrdličková Z. (2014) "Concrete governmentality: shelters and the transformations of preparedness" *Sociological Review* 62: 183–210.

Diga K. and Bracking S. (2015) *Descriptive Statistics – The developmental value of the Durban port expansion project* Research Report 2015 No. 2. University of KwaZulu-Natal. Based on the Clairwood value survey 2013 [dataset]. Version 1.0. Durban: South Africa Research Chair Initiative (Applied Poverty Reduction Assessment [producer], 2015. Durban: School of Built Environment and Development Studies [distributor], and the Leverhulme Centre for the Study of Value https://appliedpovertyreduction.files.wordpress.com/2015/01/technical-paper-2-final.pdf Accessed 21 August 2017.

Dyer J. (2014) "The consequences of Durban's proposed port development" Presentation at the University of KwaZulu-Natal's Centre for Civil Society, Durban, June 20. URL: http://ccs.ukzn.ac.za/default.asp?11,61,3,3232.

Feigenbaum T. (2011) *Measuring Anticipated Economic Impact of Dig-Out Port Development: Order of Magnitude Study*, 16 March. UrbanEcon, Durban.

Ferguson J. (2006) *Global Shadows: Africa in the Neoliberal World Order.* Duke University Press, Durham, NC.

Foucault M. (1980) "The confession of the flesh" in Gordon C. (ed.) *Power/Knowledge: Selected Interviews and Other Writings.* Pantheon Books, New York, 194–228.

Foucault M. (2003) *'Society Must Be Defended': Lectures at the Collège de France, 1975–1976.* Picador, New York.

Government of the Republic of South Africa (2014) Infrastructure Development Act, No. 21 of 2014.

Graham Muller Associates (2009a) Chapter 8 Composite Summary & Strategic Issues Final Draft March, *A Local Area Plan and Land Use Management Scheme for the Back of Port Interface Zone.* Durban.

Graham Muller Associates (2009b) "Addendum B: Status Quo Assessment" *A Local Area Plan and Land Use Management Scheme for the Back of Port Interface Zone.* Durban.

groundWork and South Durban Community Environmental Alliance (SDCEA) (2016) Media Advisory: Port Expansion Squashed – Victory for Communities and Environment. Durban.

Hallowes D. (2014) *Planning Poverty: The NDP and the Infrastructure of Destruction.* Pietermaritzburg: groundWork.

Haraway D. J. (1997) *Modest_Witness@Second_Millennium.FemaleMan_Meets_OncoMouse: Feminism and Technoscience.* Routledge, New York.

Harvey P. (2010) "Cementing relations: the materiality of roads and public spaces in provincial Peru" *Social Analysis* 54: 28–46.

Harvey P. and Knox H. (2012) "The enchantments of infrastructure" *Mobilities* 7(4): 521–536.

Hildyard N. (2016) *Licensed Larceny.* Manchester University Press, Manchester.

Khan N. (2006) "Flaws in the flow: roads and their modernity in Pakistan" *Social Text* 24: 87–113.

Larkin B. (2013) "The politics and poetics of infrastructure" *Annual Review of Anthropology* 42: 327–343.

Loftus A. and March H. (2015) "Financializing desalination: rethinking the returns of big infrastructure" *International Journal of Urban and Regional Research* 40(1): 46–61.

Marres N. (2012) *Material Participation: Technology, the Environment and Everyday Publics*. Palgrave, Basingstoke.

Mennicken A. and Sjögren E. (2015) "Valuation and calculation at the margins" *Valuation Studies* 3: 1–7.

Mrázek R. (2002) *Engineers of Happy Land: Technology and Nationalism in a Colony*. Princeton University Press, Princeton, NJ.

Nepad Business Foundation (2012) Press Release: Infrastructure Africa Business Forum 2012 Transforming Africa through Infrastructure, http://nepadbusinessfoundation.org/download/INFRASTRUCTUREAFRICABUSINESS%20FORUM2012.pdf (Accessed 21 August 2017).

Ouma S. (2014) "Situating global finance in the Land Rush Debate: a critical review" *Geoforum* 67: 162–166.

SDCEA (South Durban Community Environmental Alliance) (2014) *Comments on the Deepening, Lengthening, and Widening of Berth 203 to 205, Pier 2, Container Terminal, and Port of Durban – Amended EIA Report*. SDCEA, Durban.

Star S. L. (1999) "The ethnography of infrastructure" *American Behavioral Scientist* 43(3): 377–391.

State Capture Research Project (2017) "Betrayal of the promise: how South Africa is being stolen," http://pari.org.za/wp-content/uploads/2017/05/Betrayal-of-the-Promise-25052017.pdf (Accessed 21 August 2017).

Thévenot L. (2015) "Certifying the world: power infrastructures and practices in economies of conventional forms" in Aspers P. and Dodd N. (eds) *Re-Imagining Economic Sociology*. Oxford University Press, Oxford, 195–223.

Todorov V. (1994) *Red Square, Black Square: Organon for Revolutionary Imagination*. State University of New York Press, Albany.

Torrance M. (2009) "The rise of a global infrastructure market through relational investing" *Economic Geography* 85(1): 75–97.

Tshabalala A., Diga K. and Bracking S. (2017) Exploring the informal business sector in Clairwood, Durban, South Africa, SARCHi Technical paper number 14, http://appliedpovertyreduction.ukzn.ac.za/home/publications/ (Accessed 21 August 2017).

Tsing A. L. (2012) "On nonscalability: the living world is not amenable to precision-nested scales" *Common Knowledge* 18: 505–524.

Von Schnitzler A. (2008) "Citizenship prepaid: water, calculability, and techno-politics in South Africa" *Journal of Southern African Studies* 34: 899–917.

Part 2

Conservation

6 Bonding nature(s)?

Funds, financiers and values at the impact investing edge in environmental conservation

Sian Sullivan

Introduction

> The bond market is the great innovation that distinguishes western capitalism from previous economic systems
>
> (Climate Bonds Initiative 2010, 1)

A key element of the work of the Leverhulme Centre for the Study of Value has been to document and analyse the many different ways through which concepts and indicators of nature health and harm are becoming economised and capitalised. This exploration of the performance of economic value for units of nature health and harm *in situ* has drawn attention to the discursive, numerical-economic, material and institutional dimensions and consolidations through which new economic values for nature are articulated. 'Articulation' here is understood in two senses: as speech act utterances that shape discursive reality as understood amongst those participating in relevant speech communities (Austin 1962); and as acts of 'joining' and connection between people, organisations and practices associated with the qualification, quantification and materialisation of nature-as-natural-capital (MacDonald and Corson 2012; Corson et al. 2013; MacDonald 2013; Sullivan 2013a, 2014, forthcoming; Wilshusen and MacDonald 2017).

Our research has focused on the following cases and issues: analysis of planning negotiations and policy proposals for biodiversity offsetting at connected sites of nuclear power production (UK) and uranium extraction (Namibia) (Sullivan 2013b); the application and adjustment of technical calculative devices for the quantification of biodiversity value at sites of damage and conservation in biodiversity offsetting in England (Carver 2015; Carver and Sullivan 2017, Chapter 7 this volume; Sullivan and Hannis 2017); value struggles arising over biodiversity offsetting as a new conservation technology for mitigating harms caused by development (Sullivan and Hannis 2015); development of new balance sheet accounting practices for bringing nature as 'natural capital' into national and corporate annual accounts (Sullivan and Hannis 2017); and the ontological dimensions and politics infusing the controversial revisioning and making of 'nature' as 'natural capital' (Sullivan 2014, 2017a, 2017b, forthcoming; Sullivan and Hannis 2017). These analyses have drawn attention to the multiplicitous ways in which value(s) for nature(s) are *fabricated* rather than found (Sullivan 2017c, forthcoming). As such, they align with broader approaches in economic sociology that assert *performative* dimensions of numerical-economic devices and counting practices as they are applied to the calculation and valuation of environmental healths and harms (for example, Fourcade 2011; Ehrenstein and Muniesa 2013; Lippert 2014; Lohmann 2014; Dempsey 2016).

Drawing on review of grey literature sources, as well as attendance and participation in a series of environmental policy events in recent years (listed in Sullivan forthcoming), in this chapter I focus on new ways in which measures of the health and improvement of natures *in situ* are becoming a focus for financial concerns in 'impact investing'. Echoing notions of 'standing forests' – in which forests tend to be viewed in terms of economic values waiting to be articulated and then captured through harvest of the timber embodied in the trees that are 'standing' (as documented in Scott 1998) – I use the term '*standing natures*' to invoke a view of natures *in situ* being similarly approached as sources of new economic values that are *poised for* potential capitalisation and marketisation. This view underscores a potentially speculative form of value realisation based on assumed correspondences between natures standing now, current marketable prices, and assumptions regarding likely future scenarios that can be economised and capitalised. In particular, I set innovations we are seeing today that seek to create conserved natures *in situ* as an *asset class* for impact investing within a broader turn towards impact investing in social and development contexts. In doing so, I provide some illustrations of how conserved and restored natures *in situ* are becoming part of the bond market identified in the epigraph above as the great innovation distinguishing western capitalism from previous economic systems. In the next section I introduce the concept and practice of impact investing. I then trace some developments in the field of impact investing for the conservation, restoration and rehabilitation of terrestrial ecosystems.

What is impact investing?

Broadly speaking, impact investing (II) is a form of financing inspired by venture capitalism that promises a *dual return* on investment: returns on investment are calculated in terms of both profit and positive social, and increasingly environmental, impacts – hence *impact* investing (Chiapello and Godefroy 2017, 152). In II, the aim is to blur the boundary between philanthropic gift-giving and financial returns. Impact investing is framed increasingly as desirable as an 'innovative' source of financing for social, development and environmental concerns normally funded through public spending, because of declines in public funding – especially in the wake of the 2007–2008 financial crisis.

So-called 'Innovative Financing Mechanisms' (IFMs) are financial products and instruments that tap into funds and partners beyond commitments from conventional donors and bonds from traditional international financial institutions (Douste-Blazy 2009; Ketkar and Ratha 2009; Hurley 2012; OECD 2013). Declines in public finance associated in part with neoliberal transfers of power and assets to the private sector, coupled with crisis in multiple spheres of socio-economic and environmental life, are giving rise to a perceived need to *unlock*, i.e. to release or mobilise, funds from a buoyant private sector to finance initiatives associated in the post-World War 2 period with the public sector, at least in the industrialised north. Innovative Financing Mechanisms, including II, are thus being designed as financial instruments that will 'unlock' private sector funds. The imperative to unlock these funds is in order to 'scale up' financial resources for social, development and environmental arenas that, especially in the post-World War 2 period, have been considered the responsibility of the public sector. This scaling-up of private sector resources is considered necessary so as to fill the 'funding gap' arising as public coffers have been emptied through combinations of asset transfers, bailouts, austerity regimes and declines in tax revenue. All these phenomena are linked to the propensity for crisis that even economists of the International Monetary Fund (IMF) – one of the primary institutional architects of a global neoliberal regime[1] – now affirm are amplified by neoliberal policies (Ostry et al. 2016). Indeed, 'funding gap', 'unlock' and 'scaling-up' are terms frequently used in promotional, advocacy and technical documentation regarding II as a source of 'innovative finance' for social, development and environmental investments (for example, WWF et al. 2011; OECD 2013; Centre for Public Impact 2017: 1; Environmental Finance 2017; see review in Sullivan 2012).

Given the current plutonomic concentration of financial assets in the wealth of (ultra) high net worth individuals ((U)HNWIs) and large investors such as pension funds (Kapur et al. 2005; OXFAM 2016), finance for social, development and environmental concerns is increasingly being sought from these private sources (Mawdsley 2015). Private sector funds designed to be complementary to and interwoven with more conventional aid and public spending commitments, are thus a focus of urgent calls for a shift in global investment patterns. The intention is to generate new funds for connected endeavours in social development, the conservation of terrestrial ecosystems, carbon/climate management, and infrastructure development, including renewable energy. New financial products designed so as to source private finance for a spectrum of interconnected global challenges include: social impact bonds (Centre for Public Impact 2017; discussion in Harvie in press); green development and climate bonds[2]; and, increasingly, a variety of impact investments linked with conservation outcomes in terrestrial ecosystems or 'standing natures' (Credit Suisse et al. 2014; Huwyler et al. 2014; Credit Suisse and McKinsey Center for Business and Environment 2016; Credit Suisse et al. 2016). These financial instruments are designed to appeal to desires amongst the wealthy of society to 'do good' in an entrepreneurial spirit, and are thus also referred to as enabling 'venture philanthropy'[3] and/or 'impact entrepreneurship' (Social Impact Investment Taskforce 2014).

Impact investing constitutes a particular strand in an array of so-called IFMs for social, development and environmental agendas. As depicted in Figure 6.1, II sits between conventional finance-first investment and pure grant- or gift-giving. Chiapello and Godefroy (2017, 153) describe II as involving an *adaptation* of financial investors' practices to social and now environmental sectors, creation of a new type of investment fund, i.e. an impact investment fund, and a *redefining* of the existing funding practices of social and/or environmental organisations such that they become oriented towards and entangled with financial asset management structures. The II market is construed and constructed as a *concerned market* whose *performance* is measured on the basis of a *double return* in terms of value: 'a financial one for the investors and a social [and/or environmental] one for the public interest' (Chiapello and Godefroy 2017, 158). As a form of 'payment-by-results (PbR)' financing, II brings together socially and environmentally motivated private finance organisations with social and environmental providers. Previously public services thereby become outsourced both to private sources of finance and to private sources of provision, albeit usually with some form of publicly funded incentive or subsidy (Centre for Public Impact 2017, 1). Hailed as a revolutionary 'paradigm shift in capital market thinking' that uses private finance to leverage the value of public expenditure, II is framed by proponents as the *invisible heart* of capital markets: as a force capable of harnessing 'entrepreneurship, innovation and capital to power social [and environmental] improvement' (Social Impact Investment Taskforce 2014, 1).

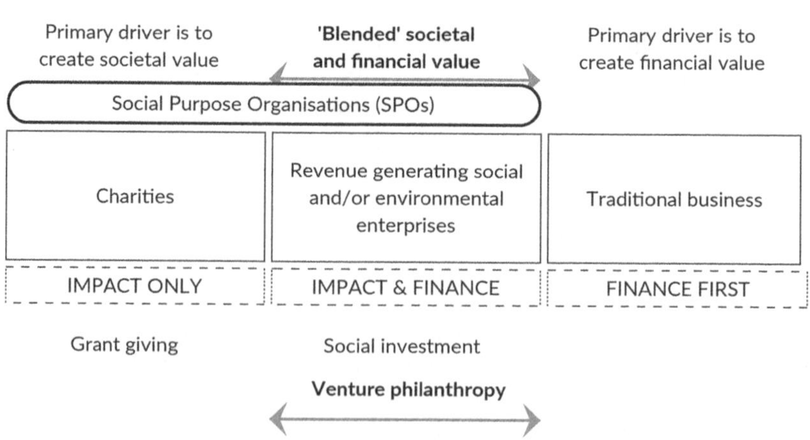

Figure 6.1 The impact investing spectrum

Source: Adapted from the European Venture Philanthropy Association (see http://evpa.eu.com/about-us/what-is-venture-philanthropy, accessed 25 July 2017). Also see Social Impact Investment Taskforce (2014, 19) and Chiapello and Godefroy (2017, 164).

Importantly, the II market is based on *promises that bind* in two main ways: investment becomes bound with the promise of positive and measurable social and/or environmental impact; simultaneously, the social and or environmental asset to be enhanced through investment becomes bound to the investment and its *promise of profit to investors.*

We can see this binding more clearly by looking at examples of how the II market is starting to play out in practice in the social development arena. The UK, for example, has seen an expansion of investment in social development under the rubric of 'Social investment' and 'Social Impact Bonds' or 'SIBs'. This expansion has been guided by a Social Impact Investment Taskforce set up by the G8 and led by a co-founder of the first venture capital fund set up in Europe in 1972 (Social Impact Investment Taskforce 2014; Chiapello and Godefroy 2017, 162). Whilst chairing the G8's Social Impact Investment Taskforce, which oversees more than half of the 30-plus SIBs currently in operation in the UK, this individual was simultaneously the chair of a social impact investment fund called Big Society Capital,[4] set up to finance social investments (Harvie in press). The UK government under the former Conservative and Liberal Democrat coalition has been instrumental in supporting policies that will grow the II market, producing a number of policy papers between 2011 and 2014[5] (for example, HM Government 2014). Since the first SIB was established in the UK in 2010, the UK now hosts the largest number (n = 32) of SIBs financing and providing social services, of a total global market in SIBs of around n = 61 (Centre for Public Impact 2017, 2). SIBs currently constitute only a very small proportion of total public service expenditure in the contexts specified. Nonetheless, II is considered indicative of 'a wider shift in the way public services are funded and delivered', especially in the developed world (Centre for Public Impact 2017, 3), with II models – specifically Development Impact Bonds (DIBs) – also being extended as financing structures for development aid spending (Centre for Global Development and Social Finance 2013).

To clarify how an SIB works, let us take an example of one UK SIB summarised in a paper by economist David Harvie (in press, and references therein) who writes that:

> The Essex SIB, launched in April 2013, was the first SIB commissioned by a local authority, Essex County Council. The bond funds a £3.1 million scheme to 'help 380 vulnerable 11–16 year olds on the edge of care or custody to stay safely at home with their families', using an intervention called Multi-Systemic Therapy (MST), provided by children's charity Action for Children. Investors include Big Society Capital [see above] and Bridges Ventures, with their £3.1m investment channelled via a special purpose [outsourced] vehicle Children's Support Services Ltd. The key outcomes metric is 'the saving in aggregate care placement days for each MST cohort, benchmarked against a historical comparison group'. The project will operate for five years, with outcomes

> payments, which will be capped at £7m extending into the eighth year. According to the [government-run] Centre for Social Impact Bonds[6], '[t]he SIB could see investors earn 8–12% annual interest on their investment', while the total saving to Essex County Council (net of outcome payments) could be £10.3million over the project's timeframe.

As this example clarifies, this 'financialization of the social' (Chiapello and Godefroy 2017, 153) is described as generating multiple wins: social impact is achieved, investors receive returns and government saves on expenditure. In addition, a secondary market seems to be emerging in connection with social II in the form of a 'social impact exchange',[7] described as 'the world's first regulated exchange dedicated to businesses and investors seeking to achieve a positive social and environmental impact through their activities'.

This broader review of II in the SIB market indicates several iterative features:

1. a funding gap is identified requiring the 'scaling-up' of resources from the private sector;
2. work begins and becomes continuous to design and create new investment products that will attract private sources of investment and to assert and stabilise new categories and boundaries that create asset classes that can attract returns-seeking investment. Through this work, new asset classes of public social development domains are identified and stabilised through combinations of professional investment and risk management skills, as well as through the design and application of standardised metrics, benchmarks, ratings, devices, etc. for the assessment, classification, categorisation and valuation of both financial and social impact returns (Chiapello and Walter 2016; Chiapello and Godefroy 2017). These activities are coupled with the significance of stabilised value judgements 'used to value not only products that are being exchanged, but also the actors that produce them' (Chiapello and Godefroy 2017, 157), contributing both to consolidations of epistemic communities around II and to the stabilising of criteria on which to base costs and prices. Linked with the bringing forth of new asset classes, and the instituting of new accounting and audit practices for both asset and impact valuation, is the emergence and consolidation of a new cadre of intermediaries offering specialist services in accounting, impact measurement, brokerage and fund management;
3. *seed financing* is frequently offered to stimulate the design and stabilising of new II pathways to 'build a marketplace' and 'unlock capital' (Chiapello and Godefroy 2017, 161). For example, in 2008, the Board of Trustees of the Rockefeller Foundation approved $38 million in support of an Impact Investing Initiative in the US for the period 2008–2011, and this is one amongst many such examples of initial funding commitments pledged to develop a market in II (discussed further below);

4 impact investing is made attractive through government support and subsidies, especially through the transfer of public funds, tax breaks and guarantees (Centre for Public Impact 2017, 14–15) coupled with competition amongst private investors for these sources of public financing. In addition, the state establishes the economic and policy context for social II by setting up appropriate regulatory and legal frameworks, and creating key supporting institutions.

The reason for going into the social II model in some detail is to clarify that models for II already exist that are now being transferred to and (re)designed for environmental concerns. This chapter thereby illustrates how calculative and valuation devices can be standardised in one value domain, and then circulated and applied to others, presumably with potential for commensurabilities between different social and environmental outcomes to also materialise. In the following section I turn to the II edge for valuing, conserving, restoring and rehabilitating 'standing natures'.

The Impact Investing Edge in Conservation Finance

> We strongly believe that natural capital is the next frontier of impact investment and both specialist skills and a critical size are required to successfully address this nascent but promising market.
>
> (Philippe Zaouati, CEO of Mirova Responsible Investing, quoted in Natixis, Mirova and Althelia Ecosphere 2017, 2)

Echoing the circumstances described above, a significant funding gap is considered to characterise environmental conservation, restoration and rehabilitation, requiring filling through IFMs, including II. Aichi Target 20, agreed in 2010 by the Conference of Parties of the Convention on Biological Diversity, thus focuses on the need to scale up financial resources for the conservation of biodiversity,[8] and an Annexe to the COP's Strategic Plan for Biodiversity in 2011–2020 speaks of the possibility for IFMs to help fill this gap.[9] Assertions of the need to scale up financial provision in the environmental arena have been accompanied by growth in proposals for how to unlock new sources of private finance so as to support investments in the conservation, restoration and rehabilitation of 'standing natures', particularly in tropical contexts in the 'global south'. Simultaneously, a number of new financing facilities have appeared to support 'conservation enterprise' via II for conservation: see, for example, Rewilding Europe Capital[10] and Conservation Capital,[11] the latter of which has a specific emphasis on Conservation Impact Investment[12] and also partners with African conservation projects through African Wildlife Capital.[13] Similarly, the ENGO The Nature Conservancy (TNC) now sells 'Conservation Notes' for a minimum investment of $25,000 over 1, 3 or 5 years, with a dual return of interest rates up to 2 per cent (depending on term selected) and environmental returns measured annually in 'acres protected, landscapes preserved, and habitat restored' (TNC 2017, online).

In what follows I document and review some moments in the emergence of proposals for II to support the conservation of standing natures *in situ*. My aim is to show how this II edge is opening and taking hold in the environmental conservation arena, some of the organisations and consolidating collaborations involved, and some of the value-framings being mobilised so as to generate both financial and conservation impacts through instruments becoming operational at this edge.

A pertinent starting point occurs in 2011. In this year a collaboration between the global ENGO the World Wide Fund for Nature (WWF), the Global Canopy Programme, the Climate Bonds Initiative and the banking/investment firms Goldman Sachs and Lombard Odier consolidated proposals for the development of 'Forest Bonds' to finance what they call 'ecological infrastructure such as tropical forests' (Kanak et al. 2011, 5). As occurs with II for social development, this collaboration proposed that public sector funds and incentives such as tax breaks be used to support private sector investment in forests, in return for government issued bonds based in part on the monetary value that could be mobilised from the 'natural capital' of tropical forests (WWF et al. 2011). The advice of these organisations was that 'the investment proposition needs to be large and liquid to attract the largest investors', and that multilateral donors might 'play an additional catalytic role by issuing a forest bond themselves and helping to pump-prime the forest bond market' (WWF et al. 2011, 5). The income for repaying the investment was proposed to derive from new sources of forest carbon revenue, generated, for example, through the United Nations REDD+ programme[14] (which received new support through the Paris Climate Agreement in 2015[15]), ecosystem service markets, sustainable timber and agriculture and taxes, all of which would arise as outcomes of the positive and measurable impacts effected through proposed private sector investments (WWF et al. 2011).

To take one source of projected value from the standing natures mentioned here, forest carbon revenue is a reference to the 'future streams of payments for expected emissions reductions' (World Bank Group 2012, 1) provided by the carbon contained in standing forests that are projected to be 'unlocked' under REDD+. The REDD+ programme encourages forests of the global south deemed to be under sustainable forest management and involving the conservation and enhancement of carbon stocks beyond a counterfactual scenario of forest degradation to become tradable in global emissions offsets markets. This process involves various (and contested) practices of quantification, calculation and monetisation, relating in particular to both documenting the quantity of carbon stored in trees left standing and the counterfactual calculation of carbon that would otherwise be lost without a REDD+ scheme in place (for details see especially Lohmann 2009, 2014; Ehrenstein and Muniesa 2013; Asiyanbi 2017).

In parallel, over the last few years new investment funds have been established to raise capital to invest in emerging markets associated with environmental conservation efforts. The Althelia Climate Fund (ACF) is prominent amongst these new II funds focusing on environmental conservation parameters. Althelia Climate Fund is a Specialised Investment Fund based in Luxembourg, established and managed by asset management platform Althelia Ecosphere – 'an impact investment manager that delivers financial returns that are fully aligned with the conservation of nature and sustainable social development'[16] – and supported by ENGO Conservation International (Kill 2016). Althelia Ecosphere invests 'in activities with the objective of providing net-positive impacts for their communities and ecosystems'.[17] Performance is measured according to a seven-pointed star of categorised 'impact themes', namely 'climate', 'ecosystems', 'species', 'livelihoods', 'inclusivity', 'sustainable enterprise' and 'fair economic returns', framed specifically in connection with the UN Sustainable Development Goals (SDGs) (Althelia Ecosphere 2016, 2017 – access to the full impact data tables is protected, requiring an investor login to the Althelia Ecosphere website). Kill (2016, 9) reports that the administrative agent and custodian bank of ACF is BNP Paribas Securities and Services, and the two major founding shareholders of the Althelia Climate Fund (as well as the major shareholders of the entity that holds a third General Partner share in ACF) previously held positions as head of Carbon Finance and Director of Environment and Forestry for BNP Paribas Corporate and Investment Banking.

In 2013 ACF was described as comprising 'a diversified portfolio of investments in Africa, Latin America and Asia that take the form of real assets (certified commodities and agricultural produce) and environmental services (verified emissions reductions and other ecosystem services)' that will deliver 'cash dividends to investors' (Althelia Ecosphere 2013, 1). At the outset the fund proclaimed an interest in potential flows of both conservation and financial value arising in particular through REDD+ and Payments for Ecosystem Services (PES) markets (Abusaid 2011). Its intention has been to create 'new environmental assets that reflect the value of natural capital',[18] with '[e]cosystem goods and services from Natural Capital' proclaimed as 'worth trillions of US dollars per year' (Althelia Ecosphere 2013: 3). Initial seed investments in the ACF from the public financial institution the European Investment Bank (EIB) amongst other investors totalled €101 million in June 2014 (Kill 2016, 2), asserted as fully invested by 2017 (Althelia Ecosphere 2017). The fund was enhanced with more than $130 million lent from the USAID in 2014,[19] assisting with guaranteeing prices to investors financing Althelia's global portfolio of investments (Kill 2016, 8). Althelia Climate Fund was paid some €4.40 million in 2014–2015 in management fees charged as a proportion of investment (Kill 2016, 4). Figure 6.2 illustrates the II institutional structure and entanglements consolidating around ACF and connected organisations.

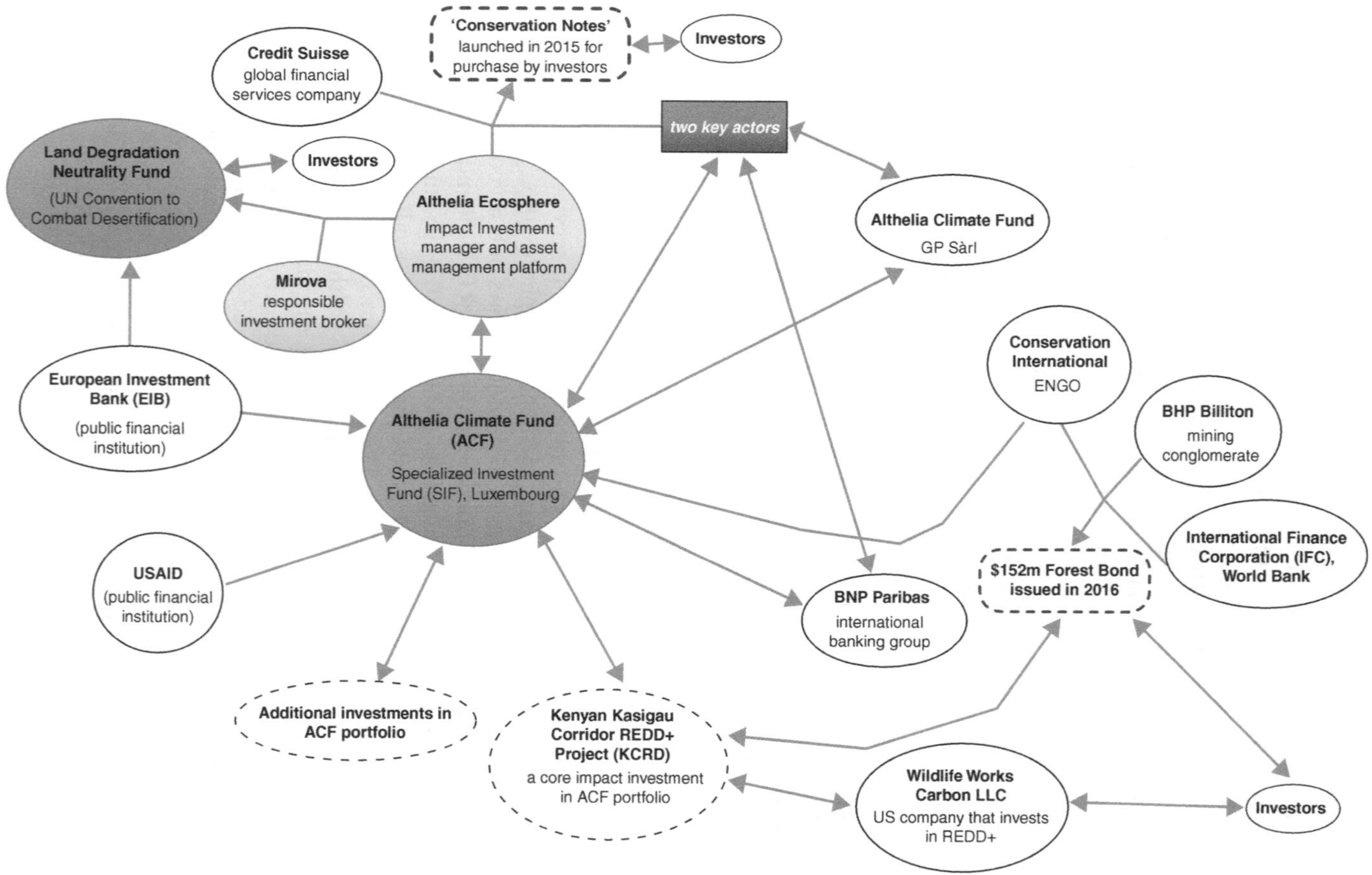

Figure 6.2 Impact investment organisational structure emerging around Althelia Climate Fund and associated organisations

Source: Derived from information reported in Kill (2016), Klopfer and Panajyan (2016), Cobley (2017: online) and https://althelia.com/ (accessed 27 July 2017).

Notes

A core project in Althelia's portfolio is the Kenyan Kasigau Corridor REDD+ Project (KCRD)[20] with recent expansions into the neighbouring 'Taita Hills Conservation and Sustainable Land Use Project'. Here, Althelia has collaborated with the US-based company Wildlife Works Carbon LLC to commit funds that invest in REDD+ and that repay investors through income arising from the sale of carbon credits from woodlands conserved and rehabilitated through the Kasigau–Taita projects. This II initiative has recently received new impetus through the release in 2016 of the first Forest Bond designed to pay its investors with both cash and/or carbon credits. Issued by the International Finance Corporation (IFC) of the World Bank and developed with mining conglomerate BHP Billiton and Conservation International, this '[i]nnovative $152 million bond to protect forests and deepen carbon-credit markets' (Klopfer and Panajyan 2016, online) represents an important first articulation between two accounting nodes of green finance architecture: the green bond market and the carbon-accounted offset market. In this case, carbon credits are purchased from the REDD+ and conservation scheme in Kenya's Kasigau Corridor, operated by the private sector company Wildlife Works LLC and amongst the portfolio of investments brokered by Althelia Climate Fund. Repayments in certified carbon credits can be used either to offset corporate carbon emissions or sold on carbon markets (Klopfer and Panajyan 2016, online). The 5-year bond is designed to scale up private sector climate change finance and conservation liquidity, but nonetheless exists in a context of concern regarding local socio-economic impacts of Kasigau offset provision (Chomba et al. 2016; Kill 2016). The bond has been sold to major global institutional investors, including the pension funds California State Teachers' Retirement System (CalSTRS) and Teachers Insurance and Annuity Association of America (TIAA), the impact investment fund Treehouse Investments LLC, and the insurance company QBE (Klopfer and Panajyan 2016: online). As well as preventing deforestation, proceeds from the bond are intended to support private sector development and particularly IFC's private sector lending in emerging environmental markets (Klopfer and Panajyan 2016, online).

Forest Bonds are also being established that focus specifically on Indonesian forests. In April 2015, ADM Capital,[21] an investment manager seeking long-term capital appreciation through opportunities in Asia and Eastern Europe, collaborated with the ENGO Flora and Fauna International,[22] to launch a $1 billion bond programme in 'Rainforest Impact Bonds' as a finance mechanism for tropical forest conservation that stimulates green economic growth (ADM Capital 2015). This initiative has recently been boosted by a new grant to support the design of a Tropical Landscapes Finance Facility (TLFF) and Tropical Landscapes Bond (TLB), developed in partnership with the United Nations Environment Programme (UNEP), the International Center for Research in Agroforestry (ICRAF), and (again) the bank BNP Paribas (Genasci 2017). As described above, these bond structures are designed in connection with sovereign aid commitments from developed countries to stem global climate change by reducing forest carbon emissions through deforestation and habitat degradation. The flow of repayments to investors in Rainforest Impact Bonds is thus projected to derive from newly commodified and marketable carbon values in tropical forests whose value has been made legible in part via sovereign aid commitments derived from public monies.

These forest and rainforest bonds as instruments for supporting II in the conservation of standing natures are themselves part and parcel of proliferating proposals for scaling-up conservation investments from institutional investors and (ultra) high net worth individuals ((U)HNWIs), i.e. the super-super-rich, through financial products linked with emerging or predicted conservation markets (Huwyler et al. 2014, online; also Credit Suisse et al. 2014; Credit Suisse and McKinsey Center for Business and Environment 2016). In these new impact-related conservation finance structures, investor risk is projected to be reduced through mobilising such newly eligible-leverageable assets and the 'land or usage rights' from which they derive as underlying collateral (see, for example, Credit Suisse and McKinsey Centre for Business and Environment 2016, 17). Credit Suisse is emerging as a key financial institution in mobilising debt-based conservation finance, having also worked with Althelia Ecosphere (see above) to sell 'Conservation Notes' in 2015,[23] raising $14 million from around 50 HNWI investors for conservation projects invested in via Althelia Climate Fund's portfolio (Kill 2016, 7). Credit Suisse envisage and propose the matching of portfolios of conservation projects with a portfolio of investments, as shown in Figure 6.3, a structure that makes it increasingly hard to link any particular investment with any particular conservation project or outcome.

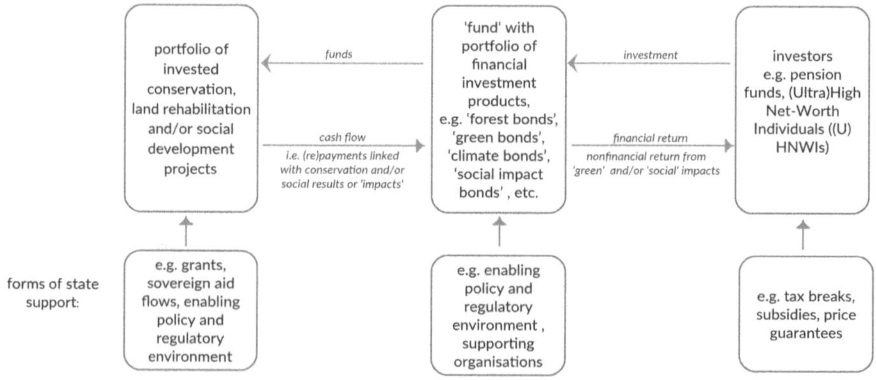

Figure 6.3 Schematic representation of impact investing finance for environmental and social impacts and associated forms of state support

Source: Drawing on Credit Suisse, WWF and McKinsey&Company (2014, 11), ADM Capital (2015) and Credit Suisse and McKinsey Center for Business and Environment (2016, 9).

The above examples focus on preventing the degradation of high-value 'standing natures'. A different but related emerging aspect of II for conservation focuses on financing structures for the rehabilitation of land considered degraded. Once again, 'large amounts of financial resources' are asserted as required to fill the funding gap understood as a barrier to meeting the UN SDG target (15.3) of 'a land degradation neutral world' by 2030, and are envisaged to be provided by '[n]ew financial instruments and intermediaries, as well as enabling conditions, ... to catalyze private capital around SDG implementation'.[24] One response to this nexus of constraints and possibilities, and as outlined by the UN Convention to Combat Desertification (UNCCD), is the establishment of an independent Impact Investment Fund for Land Degradation Neutrality (LDN), launched in 2017 as 'a first-of-its-kind investment vehicle leveraging public money to raise private capital for sustainable land management and land restoration projects worldwide'.[25]

Echoing the structures outlined above, the LDN Fund will be designed and managed by Mirova, a 'responsible investment' affiliate of French investment company Natixis Asset Management[26] managing 'nearly €7bn in green equity, bond, impact and infrastructure funds', and seeking to raise €1 billion for the LDN Fund (Cobley 2017, online). Private finance is to be complemented by seed finance and support from public financial institutions including the EIB.[27] This structure means that 'state backers [will] make up around 30% of the fund's investor base, under an arrangement that would see them take the first hit if any investments go sour' thereby lowering 'risk for private-sector investors like pension funds, ... the idea being that private investors would benefit from returns of around 8% a year' (Cobley 2017, online). In a bid to become 'the European leader in natural capital investing' (Natixis et al. 2017, 1), Mirova, under contract with UNCCD, is negotiating a majority equity interest in Althelia Ecosphere (see above), through which it will seek to build and manage the LDN Fund as 'a public–private investment fund to regenerate and manage farmland around the world – at a profit' (Cobley 2017, online). The LDN Fund is described as 'being structured as a layered fund, designed as a public–private partnership for blended finance' to 'complement and scale up existing financial instruments and funds for sustainable land management and rehabilitation by providing financing that would not otherwise be available in the market'.[28] These developments are thereby reframing and approaching land considered degraded as a new asset class suitable for investment (WBCSD 2015).

To summarise, the examples outlined in this section repeat and consolidate the II structures considered in the previous section. They take shape in contexts of oft-stated concerns regarding the gaps in available finance for public expenditures. Reflecting broader interest in financial instruments that 'blend' public and private provision they frequently draw on seed funding and other guarantees from public financial institutions, and are supported by enabling regulatory and policy structures. They involve the vesting of skills, control and fees in intermediaries located in finance, thereby entangling solutions for environmental concerns with expertise in financial asset creation and management. In the following and concluding section I consider some structural implications of this II edge in financing the conservation, restoration and rehabilitation of standing natures.

Conclusion: 'You know what I mean by a bond? Something that binds'[29]

> Institutional investors are actually very good at coming to terms with new sectors and making investments. But the thing that drives them to do that is returns. If there is a way things can be structured to get higher returns, the market will work really well.
>
> (Rupesh Madlani, CEO Global Sustainable Capital Management, in Environmental Finance 2017, online)

> Put together, these resources would provide the strategic vision for growth, strong investment processes, and support functions needed to scale up this natural capital investing business.
>
> (Natixis, Mirova and Althelia Ecosphere 2017, 1)

The proposals outlined above for binding standing natures with financial domains through II aim to transform the supply of private capital to social, development and ecological enterprise with measurable impact, i.e. that better evidences the value (for money) that has been created. They require the creation of new asset classes of conserved, restored and rehabilitated 'standing natures' that can attract or 'unlock' new sources of private capital investment so as to fill identified funding gaps in possibilities for social and environmental improvement. This, then, is a financing approach in which the qualification of 'standing natures' as 'natural capital' *in situ*, coupled with quantification through (ac)counting and valuation technologies, is permitting translations of projected natural capital quantities into financially material, i.e. leverageable, forms of value.

Impact investing for the conservation, restoration and rehabilitation of standing natures is a new edge in the world of conservation finance. It is too early to say exactly what social, economic and material effects these innovations will manifest, or how much money will flow, to where and to whom. Some commentators have asserted the nascent, halting and faltering nature of this new investment field (for example, Dempsey and Suarez 2016; Dempsey 2017). As Asiyanbi (2017, 1) writes, however, while 'material processes appear thin, sluggish, fractured, hybridised or stalled in practice', the emergent and expanding green projects outlined above can simultaneously be understood as 'durable processes of becoming' and as consolidating, rather than retreating, 'markets-in-the-making'. Indeed, several of the examples shared here illustrate a keen appetite in ensuring success in this new edge in conservation finance on the part of their architects in public financial institutions, private intermediary organisations and 'on-the-ground' suppliers of carbon and ecosystem value. This appetite is demonstrated by new seed investments, new alliances between intermediaries that may assist with scaling up investment portfolios, and consolidated policy support at national and international levels. All these dimensions are 'organising actions' exhibiting a neoliberal tendency towards the 'truth regime of the market' (Foucault 2008(1979)) wherein interventions in environmental management are rendered increasingly intelligible to market and financial logics (Sullivan 2013a; also Asiyanbi 2017, 5).

At the same time, financialisation can be thought of in multiple ways that are not about financial flows only (for example, Sullivan 2013a; Chiapello and Godefroy 2017). Financialisation is of course about volumes and directions of money that both move and accumulate via financial instruments and investments. But it is also about making and valuing new aspects of the world in the image of finance through the creation of newly (ac)countable asset classes that perhaps close off and *devalue* other(ed) and prior ways of knowing social and natural worlds. It is about creating opportunities and communities aligned around the fizz of potential and promise as new asset classes are fabricated in the world. It is about framing and taming social and natural worlds as (pump) primed for (ac)counting, acquisition and accumulation, as occurs in the creation of all speculative investment bubbles in which there is money to be made precisely because such opportunities may be unsustainable.

This chapter has traced some ways in which places and spaces reframed and revalued as asset classes are becoming further incorporated into financial(ised) value-frames, as testified by the brokers, intermediaries and investments funds for conservation finance outlined above. It is too soon to see the extent to which these proposals, policies and products will take hold, how much money will thereby flow either towards conservation investments or towards investors, or what conservation impacts may manifest in the long term and with what social effects.

This is an edge that is moving fast, but it is also an edge animated by the promise of financial(ised) returns of value and the sustenance of capital(ist) political economic structures. These structures tend to distribute income unequally, pulling towards accumulation, speculation and inequity. In II for conservation finance, they may additionally intensify processes whereby standing natural capitals in southern countries become bound to outside investors in ways that may influence sovereignty over those resources, as well as increasing the indebtedness of the thus-invested contexts. Importantly, then, if these public–private investments in public social and environmental spheres are to elicit significant public understanding and confidence they will require impact measures, data tables and reporting to be publicly available for review and debate. This is an edge ripe for detailed empirical, comparative and independent case research, especially to clarify evidence for positive social and/or environmental impacts for those living in tropical ecosystems.

Notes

1 In particular through the coalescence of globalising political and economic policies that flowed from the Washington Consensus drawn up in 1989 by economic advisers to the major international financial institutions (primarily the World Bank and the IMF). This global(ising) economic framework is complex in structure and effects but has included: deregulation of international finance flows; protection of business interests, in part via the establishment of so-called 'free trade' regions; and structural adjustment and austerity programmes to which 'developing' countries, and latterly post-crisis European countries such as Greece, have been expected to agree and adhere in order to access donor funds for development purposes. These structural adjustment and austerity programmes introduce a range of conditionalities, frequently oriented towards the opening up of 'southern' markets and utilities to international business, and a transformed role for the state so as to permit further trade and donor permeability in such contexts. Major international non-governmental organisations (INGOs) and environmental non-governmental organisations (ENGOs) mushroomed in the wake of the Washington Consensus to fill the facilitating and implementation vacuum left in circumstances of weakened states and public sectors (Foucault 2008(1979)); Hulme and Edwards 1997; Stiglitz 2002; Chapin 2004). The ways in which the combination of such structuring processes have vested sovereignty in locations beyond the nation state was famously termed *Empire* by Hardt and Negri (2000). See discussion in Sullivan (2006).
2 For an up-to-date list of bonds self-labelled as 'green' see https://www.climatebonds.net/cbi/pub/data/bonds (Accessed 25 July 2017).
3 See, for example, the European Venture Philanthropy Association at http://evpa.eu.com/ (Accessed 25 July 2017).
4 https://www.bigsocietycapital.com/ (Accessed 25 July 2017).
5 At https://www.gov.uk/government/publications/growing-the-social-investment-market-a-vision-and-strategy (Accessed 25 July 2017).
6 https://data.gov.uk/sib_knowledge_box/home (Accessed 26 July 2017).
7 http://socialstockexchange.com/ (Accessed 25 July 2017).
8 See https://www.cbd.int/sp/targets/ (Accessed 26 July 2017).
9 See www.cbd.int/sp/ (Accessed 26 July 2017).
10 https://www.rewildingeurope.com/rewilding-europe-capital/ (Accessed 26 July 2017).
11 www.conservation-capital.com/ (Accessed 26 July 2017).
12 www.conservation-capital.com/financing-conservation-ent/ (Accessed 26 July 2017).
13 www.conservation-capital.com/investment-awc/ (Accessed 26 July 2017).
14 The UN programme for Reducing Emissions from Deforestation and Forest Degradation in Developing Countries, www.un-redd.org/, the programmatic implementation of which is coordinated in particular by the Food and Agriculture Organisation (FAO), the UN Development Programme (UNDP) and the World Bank's Forest Carbon Partnership Facility (FCPF) (Kill pers. comm. 6 August 2017).
15 United Nations Framework Convention on Climate Change (UNFCCC) Paris Agreement 2015, Article 4.1 consolidates an approach to global carbon management that emphasises an *aggregate* or *net* 'balance between anthropogenic emissions by sources and removals by sinks of greenhouse gases' thereby and permitting global carbon management through offsetting, including through purchase of tropical forest carbon.
16 https://althelia.com/ (Accessed 26 July 2017).
17 https://althelia.com/our-approach/our-impacts-monitoring/ (Accessed 26 July 2017).

18 https://althelia.com/our-approach/ (Accessed 25 July 2017).
19 https://althelia.com/2014/05/29/usaid-announce-new-partnership-with-althelia/ (Accessed 25 July 2017).
20 Seewww.coderedd.org/redd-project-devs/wildlife-works-carbon-kasigau-corridor/ (Accessed 25 July 2017).
21 http://admcapital.com/ (Accessed 26 July 2017).
22 www.fauna-flora.org/initiatives/innovative-finance-for-conservation/ (Accessed 26 July 2017).
23 See https://althelia.com/2015/01/14/althelia-ecosphere-joins-forces-credit-suisse-launching-nature-conservation-notes-accelerate-private-sector-conservation-finance-preservation-ecosystems/ (accessed 31 July 2017).
24 See www.un.org/sustainabledevelopment/biodiversity/ and www2.unccd.int/actions/impact-investment-fund-land-degradation-neutrality (Both accessed 31 July 2017).
25 www2.unccd.int/actions/impact-investment-fund-land-degradation-neutrality (Accessed 31 July 2017). Thank you to James Fairhead for drawing my attention to this fund.
26 https://www.ngam.natixis.com/global/1250194644819/Home (Accessed 31 July 2017).
27 www2.unccd.int/actions/impact-investment-fund-land-degradation-neutrality (Accessed 31 July 2017).
28 www2.unccd.int/actions/impact-investment-fund-land-degradation-neutrality (Accessed 31 July 2017).
29 From Pullman (1997, 187).

References

Abusaid L.E. (2011) *Private Market Finance Options for REDD+ Activities*. Climate Markets and Investment Association. URL: http://forestclimatecenter.org/files/2011-11-27%20Private%20Finance%20Market%20Options%20for%20REDD+%20Activities.pdf (Accessed 21 May 2016 – no longer available online).

ADM Capital (2015) *Innovative Rainforest Bond Structure Unveiled at Indonesia Tropical Landscapes Summit*. URL: www.marketwired.com/press-release/innovative-rainforest-bond-structure-unveiled-indonesia-tropical-landscapes-summit-2013842.htm (Accessed 21 May 2016).

Althelia Ecosphere (2013) Althelia Ecosphere completes first closing for Althelia Climate Fund with over EUR60 million (US$80m) raised, Press release. URL: www.ecospherecapital.com/wp-content/uploads/2011/11/ACFclose_PR_FINAL.pdf (Accessed 1 August 2017).

Althelia Ecosphere (2016) *Impact Report 2016: Aligning Economy with Ecology*. URL: https://althelia.com/wp-content/uploads/2016/07/Althelia-Impact-Report-2016_FINAL.pdf (Accessed 26 July 2017).

Althelia Ecosphere (2017) *Impact Report 2017: Aligning Economy with Ecology*. URL: https://althelia.com/wp-content/uploads/2017/07/Althelia_impact-report_2017.pdf (Accessed 1 August 2017).

Asiyanbi A.P. (2017) "Financialisation in the green economy: material connections, markets-in-the-making and Foucauldian organising actions" *Environment and Planning A* doi:10.1177/0308518X17708787

Austin J.L. (1962) *How to Do Things with Words*. Oxford University Press, London

Carver L. (2015) "Measuring the value of what? An ethnographic account of the transformation of 'Nature' under the DEFRA biodiversity offsetting metric" *LCSV*

Working Paper 11. URL: http://thestudyofvalue.org/wp-content/uploads/2015/02/WP11-Carver-2015-Measuring-whatvalue.pdf (Accessed 1 August 2017).

Carver L. and Sullivan S. (2017) "How economic contexts shape calculations of yield in biodiversity offsetting" *Conservation Biology* 31(5): 1053–1065.

Centre for Global Development and Social Finance (2013) *Investing in Social Outcomes: Development Impact Bonds.* URL: https://www.cgdev.org/sites/default/files/investing-in-social-outcomes-development-impact-bonds.pdf (Accessed 31 July 2017).

Centre for Public Impact (2017) *Social Impact Bonds: An Overview of the Global Market for Commissioners and Policymakers.* URL: http://socialspider.com/wp-content/uploads/2017/04/SS_SocialImpactReport_4.0.pdf (Accessed 25 July 2017).

Chapin M. (2004) "A challenge to conservationists" *World Watch Magazine* November/December: 16–31.

Chiapello E. and Godefroy G. (2017) "The dual function of judgment devices: why does the plurality of market classifications matter?" *Historical Social Research* 42(1): 152–188.

Chiapello E. and Walter C. (2016) "The three ages of financial quantification: a conventionalist approach to the financiers' metrology" *Historical Social Research* 41(2): 155–177.

Chomba S., Kariuki J., Lund J.F. and Sinclair F. (2016) "Roots of inequity: how the implementation of REDD+ reinforces past injustices" *Land Use Policy* 50: 202–213.

Climate Bonds Initiative (2010) *Climate Bonds Can Fund the Rapid Transition to a Low-carbon Economy.* URL: https://www.climatebonds.net/files/uploads/2010/06/4-pager_climatebonds.pdf (Accessed 25 July 2017).

Cobley M. (2017) *Green French Manager in Tie-up Talks with Desert-busters.* URL: https://www.fnlondon.com/articles/french-green-manager-in-tie-up-talks-with-desert-busting-london-firm-20170629 (Accessed 31 July 2017).

Corson C., Macdonald K.I. and Neimark B. (2013) "Grabbing 'green': markets, environmental governance and the materialization of natural capital" *Human Geography* 6(1): 1–15.

Credit Suisse, WWF and McKinsey&Company (2014) *Conservation Finance: Moving Beyond Donor Funding to an Investor Driven Approach.* Online: https://www.cbd.int/financial/privatesector/g-private-wwf.pdf (Accessed 21 May 2016).

Credit Suisse, Climate Bonds Initiative and Clarmondial (2016) *Levering Ecosystems: A Business-focused Perspective on How Debt Supports Investments in Ecosystem Services.* URL: https://is.gd/6Sp8aH (Accessed April 2017).

Credit Suisse and McKinsey Center for Business and Environment (2016) *Conservation Finance. From Niche to Mainstream: The Building of an Institutional Asset Class.* URL: https://www.credit-suisse.com/media/assets/corporate/docs/about-us/responsibility/banking/conservation-finance-en.pdf (Accessed 21 May 2016).

Dempsey J. (2016) *Enterprising Nature: Economics, Markets, and Finance in Global Biodiversity Politics.* Wiley Blackwell, Oxford.

Dempsey J. (2017) "The financialization of nature conservation?" in Christophers B., Leyshon A. and Mann G. (eds) *Money and Finance after the Crisis: Critical Thinking for Uncertain Times.* Wiley, Chichester.

Dempsey J. and Suarez D.C. (2016) "Arrested development? The promises and paradoxes of 'selling nature to save it'" *Annals of the American Association of Geographers* 106(3): 653–671.

Douste-Blazy D. (2009) *Innovative Financing for Development: The I-8 Group Leading Innovative Financing for Equity (L.I.F.E.).* United Nations, New York.

Ehrenstein V. and Muniesa F. (2013) "The conditional sink: counterfactual display in the valuation of a carbon offsetting restoration project" *Valuation Studies* 1(2): 161–188.

Environmental Finance (2017) *Scaling Up Environmental Finance.* URL: www.imf.org/external/pubs/ft/fandd/2016/06/ostry.htm (accessed 25 July 2017).

Foucault M. (2008(1979)) *The Birth of Biopolitics: Lectures at the Collège de France 1978–1979* (trans. G. Burchell). Palgrave Macmillan, Basingstoke.

Fourcade M. (2011) "Cents and sensibility: economic valuation and the nature of 'Nature'" *American Journal of Sociology* 116(6): 1721–1777.

Genasci L. (2017) *ADM Capital Foundation Receive Convergence Grant for Design of Innovative Financing Facility for Indonesia.* URL: https://genascihk.com/2017/01/19/adm-capital-foundation-receive-convergence-grant-for-design-of-innovative-financing-facility-for-indonesia/ (Accessed 24 July 2017).

Hardt M. and Negri A. (2000) *Empire.* Harvard University Press, Cambridge, MA.

Harvie D. (in press) "(Big) society and (market) discipline: the financialisation of social reproduction" *Historical Materialism.*

HM Government (2014) *Growing the Social Investment Market: 2014 Progress Update.* URL: https://www.gov.uk/government/uploads/system/uploads/attachment_data/file/321483/2014_Social_Investment_Strategy.pdf (Accessed 25 July 2017).

Hulme D. and Edwards M. (eds) (1997) *NGOs, States and Donors: Too Close for Comfort?* Macmillan, London.

Hurley G. (2012) *Innovative Financing for Development: A New Model for Development Finance?* United Nations Development Programme (UNDP), New York.

Huwyler F., Kaeppeli J., Serafimova J., Swanson K. and Tobin J. (2014) "Making conservation finance investable" *Stanford Social Innovation Review* 14 January. URL: http://ssir.org/up_for_debate/article/making_conservation_finance_investable (Accessed 21 May 2016).

Kanak D.P., Mitchell A.W. and Kidney S. (2011) "Scaling up forest finance: statement from the workshop hosts" in WWF (WWF Forest and Climate Initiative), GCP (Global Canopy Programme), Climate Bonds Initiative, Goldman Sachs and Lombard Odier *Unlocking Forest Bonds: A High-Level Workshop on Innovative Finance for Tropical Forests.* URL: http://globalcanopy.org/publications/unlocking-forest-bonds (Accessed 25 July 2017).

Kapur A., Macleod N. and Singh N. (2005) *Plutonomy: Buying Luxury, Explaining Global Imbalances.* Citigroup Global Markets Inc., New York.

Ketkar S. and Ratha D. (2009) *Innovative Financing for Development.* World Bank, Washington, DC.

Kill J. (2016) *The Kasigau Corridor REDD+ Project in Kenya: A Crash Dive for Althelia Climate Fund.* Re:Common and Counter Balance. URL: www.counter-balance.org/wp-content/uploads/2017/02/The-Kasigau-Corridor-REDD_Kenya.pdf (Accessed 25 July 2017).

Klopfer A. and Panajyan S. (2016) IFC issues innovative $152 million bond to protect forests and deepen carbon-credit markets. URL: http://ifcextapps.ifc.org/ifcext/pressroom/ifcpressroom.nsf/0/594A016A78A7B14E8525805D00461397 (Accessed 25 July 2017).

Lippert I. (2014) "Environment as datascape: enacting emission realities in corporate carbon accounting" *Geoforum* 66: 126–135.

Lohmann L. (2009) "Toward a different debate in environmental accounting: the cases of carbon and cost–benefit" *Account, Organizations and Society* 34: 499–534.

Lohmann L. (2014) "Performative equations and neoliberal commodification: the case of climate" in Büscher B., Dressler W. and Fletcher R. (eds) *Nature^{TM} Inc.: Environmental Conservation in the Neoliberal Age.* Arizona University Press, Tucson.

MacDonald K.M. (2013) "Grabbing green: cynical reason, instrumental ethics and the production of 'The Green Economy'" *Human Geography* 6(1): 46–63.

MacDonald K.M. and Corson C. (2012) "'TEEB Begins Now': a virtual moment in the production of natural capital" *Development and Change* 43(1): 159–194.

Mawdsley E. (2015) "DFID, the private sector and the re-centring of an economic growth agenda in international development" *Global Society* 29(3): 339–358.

Natixis, Mirova and Althelia Ecosphere (2017) Natixis announces Mirova has entered into exclusive negotiations with Althelia Ecosphere to become the European leader in natural capital investing. URL: www.mirova.com/Content/Files/Mirova/NEWS/2017%2006_PR%20Mirova_Project%20of%20acquisition%20of%20Althelia_Natixis%20EN_Final.pdf (Accessed 1 August 2017).

OECD (2013) *Scaling-up Finance Mechanisms for Biodiversity.* OECD Publishing, Paris.

Ostry J.D., Loungani P. and Furceri D. (2016) "Neoliberalism: oversold?" *Finance and Development* 3(2) URL: www.imf.org/external/pubs/ft/fandd/2016/06/ostry.htm (Accessed 26 July 2017).

OXFAM (2016) *An Economy for the 1%: How Privilege and Power in the Economy Drive Extreme Inequality and How This Can Be Stopped.* URL: https://www.oxfam.org/en/research/economy-1 (Accessed 25 July 2017).

Pullman P. (1997) *The Subtle Knife.* Scholastic UK, London.

Scott J. (1998) *Seeing Like a State: How Certain Schemes to Improve the Human Condition Have Failed.* Yale University Press, New Haven, CT.

Social Impact Investment Task Force (2014) *Impact Investment: The Invisible Heart of Markets.* URL: www.ietp.com/sites/default/files/Impact%20Investment%20Report%20FINAL.pdf (Accessed 31 July 2017).

Stiglitz J. (2002) *Globalization and its Discontents.* Allen Lane, London.

Sullivan S. (2006) "The elephant in the room? Problematizing 'new' (neoliberal) biodiversity conservation" *Forum for Development Studies* 33(1): 105–135.

Sullivan S. (2012) *Financialisation, Biodiversity Conservation and Equity: Some Currents and Concerns.* Third World Network, Environment and Development Series 16, Penang Malaysia.

Sullivan S. (2013a) "Banking nature? The spectacular financialisation of environmental conservation" *Antipode* 45(1): 198–217.

Sullivan S. (2013b) "After the green rush? Biodiversity offsets, uranium power and the 'calculus of casualties' in greening growth" *Human Geography* 6(1): 80–101.

Sullivan S. (2014) "The natural capital myth; or will accounting save the world? Preliminary thoughts on nature, finance and values" *LCSV Working Paper* 3. URL: http://thestudyofvalue.org/wp-content/uploads/2013/11/WP3-Sullivan-2014-Natural-Capital-Myth.pdf (Accessed 1 August 2017).

Sullivan S. (2017a) "On 'natural capital', 'fairy-tales' and ideology" *Development and Change* 48(2): 397–423.

Sullivan S. (2017b) "'What's ontology got to do with it?' On nature and knowledge in a political ecology of 'the green economy'" *Journal of Political Ecology* 24: 217–242.

Sullivan S. (2017c) "Noting some effects of fabricating 'nature' as 'natural capital'" *Ecological Citizen* 1: 65–73.

Sullivan S. (forthcoming) "Making nature investable: from legibility to leverageability in the making of natural capital" *Science and Technology Studies*. https://sciencetechnologystudies.journal.fi/forthcoming/article/58040/27301

Sullivan S. and Hannis M. (2015) "Nets and frames, losses and gains: value struggles in engagements with biodiversity offsetting policy in England" *Ecosystem Services* 15: 162–173.

Sullivan S. and Hannis M. (2017) "'Mathematics maybe, but not money': on balance sheets, numbers and nature in ecological accounting" *Accounting, Auditing and Accountability Journal* 30(7): 1459–1480.

TNC (The Nature Conservancy) (2017) *Conservation Note*. URL: https://www.nature.org/about-us/conservation-note-fact-sheet-2016.pdf (Accessed 31 July 2017).

WBCSD (World Business Council for Sustainable Development) (2015) *Land Degradation Neutrality: Issue Brief for Business*. URL: www.wbcsdservers.org/web/org/Land_Degradation_Neutrality_Issue_Brief.pdf (Accessed 31 July 2017).

Wilshusen P.R. and MacDonald K.I. (2017) "Fields of green: corporate sustainability and the production of economistic environmental governance" *Environment and Planning A* 49(8): 1824–1845.

World Bank Group (2012) *Modalities and Procedures for Financing Results-Based Actions on REDD+*. Submission to the UNFCCC Secretariat. URL: http://unfccc.int/resource/docs/2012/smsn/igo/71.pdf (Accessed 27 July 2017).

WWF (WWF Forest and Climate Initiative), GCP (Global Canopy Programme), Climate Bonds Initiative, Goldman Sachs and Lombard Odier (2011) *Unlocking Forest Bonds: A High-Level Workshop on Innovative Finance for Tropical Forests*. URL: www.globalcanopy.org/materials/unlocking-forest-bonds (Accessed 21 May 2016).

7 Creating conservation values under DEFRA's biodiversity offsetting pilot and the pragmatics of using a calculative device

Louise Emily Carver and Sian Sullivan

Introduction

In recent years, the mitigation practice of biodiversity offsetting (BDO) has proliferated globally as a means of securing biodiversity compensation from impacts associated with infrastructure development (for reviews see Quintero and Mathur 2011; Benabou 2014). BDO seeks to secure the 'no net loss' and preferably a 'net gain' of biodiversity through the application of valuation technologies. Such technologies (that here, after Callon and Muniesa 2005, are labelled 'calculative devices') measure biodiversity values at sites of *development impact*, to secure compensation through equivalent or additional biodiversity values at sites of conservation *investment*. BDO is distinguished from other types of ecological compensation in English planning by taking a position that gains are measurable via metrics and proxy scores sustained over specific time periods (DEFRA 2013). The UK pioneered this approach in England in 2011, and conducted a pilot study between 2012 and 2014 to test it out with six local planning authorities across the country (DEFRA 2011b).

Preceded by wetland mitigation banking and species banking in the United States from the 1970s and 1990s respectively (Fox and Nino-Murcia 2005; Carroll et al. 2008; Robertson and Hayden 2008; Pawliczek and Sullivan 2011), BDO in England is embedded in a wider discourse concerned with revealing and 'securing the value' of biodiversity and ecosystems (DEFRA 2011a; TEEB 2010; Helm 2014, 2015). Indeed, the UK government initiated its BDO pilot study as part of this broader effort and specifically in relation to rethinking conservation policy in terms of its contribution to growing the 'green economy' (DEFRA 2011a). In DEFRA's 2011 Natural Environment White Paper announcing the BDO pilot, the Secretary of State at the time captured this sentiment thus:

> government and society need to account better for the value of nature, particularly the services and resources it provides... Valuing nature properly holds the key to a green and growing economy, one which invests in nature – not just for us but for our children's children.
>
> (Rt. Hon. Caroline Spelman MP, DEFRA 2011a, 2)

This agenda and its call to 'account better for the value of nature' implies that the ('green') economic values of biodiversity have been present all along, waiting to be recognised, accounted for and captured so as to ensure green economic growth. The interpretive starting point for this chapter instead takes valuation to be a social practice through which economic values are fabricated via the consolidation of discursive, calculative and institutionalised practices (Helgesson and Muniesa 2013; Sullivan 2017, forthcoming). As such we are concerned with processes of economic performativity that contribute to and format the reality they describe, rather than simply discovering it (Callon 1998). Economic performativity occurs through active practices of economisation and marketisation, through which behaviours, institutions and devices actively work to qualify and constitute things as 'economic' that previously have been located beyond (or 'external to') economised and marketised values (Çalişkan and Callon 2009, 2010). In this chapter we are interested in the ways in which economic qualifications can transform entities (biodiversity habitats) into market goods through application of calculative devices that facilitate 'the delimitation of the boundary between goods included in the space of market calculation and those that [are] excluded, (Callon and Muniesa 2005, 14; also Muniesa et al. 2007).

Consequently, the focus of this chapter is the 'DEFRA biodiversity offsetting metric' as a calculative device (see especially DEFRA 2012a), which in application enacts biodiversity values as numerical surrogates that produce a new entity called a 'biodiversity unit'. The biodiversity unit forms the currency of BDO as an emerging market instrument in biodiversity conservation and thereby permits the potential trade of biodiversity 'value' between developers and offset providers. As well as producing these economic values, we argue that the calculative device exemplifies practices of commensuration defined as 'the transformation of different qualities into a common metric' (Espeland and Stevens 1998, 2), and in so doing is a marketisation process involving the creation of tradable entities.

BDO is a relatively new conservation technology whose recency in application means that detailed case histories of offset implementation and actual (as opposed to projected) conservation outcomes remain limited. To support debate regarding the conservation effectiveness of BDO more generally, case study detail is important in understanding particularly how the calculative devices designed for BDO are deployed in practice. In particular, our empirical engagements trace the processes of economic value-making in conservation policy and practice and illustrate how conflicts and tensions identified within environmental markets more generally (Salzman and Ruhl 2000) are resolved by actors *in situ*.

This chapter reviews existing studies of BDO, focusing on the development and application of metrics that create seemingly comparable measures of biodiversity to demonstrate measurable gains – or 'yields' in biodiversity unit value. The BDO pilot established by DEFRA between 2012 and 2014 is then introduced and one of its key elements – namely testing through application, the calculative device of BDO technical metric – is described in detail. Findings from a 24-month field research engagement in BDO in England are then presented, to document the unfolding of a specific BDO offsets contract (also see Carver and Sullivan 2017), set within a broader dataset focusing on six DEFRA BDO pilot sites in England (Carver 2017).

We close by assessing the proposal that application of the DEFRA technical metric is able to standardise assessments of biodiversity value so as to offer robust numerical foundations on which to base planning in calculating biodiversity value for development and offset sites in England. Given that economisation is a social process, it is also therefore a political one. We illustrate through our case study that, as predicted in theory (cf. Walker et al. 2009), conservation policy decisions based on valuation and calculative (market) devices conceal and obfuscate the political character of such valuations. We argue that the calculations within biodiversity offsetting in England thus tend to reflect prior configurations of power and existing political priorities for economic over ecological values. This chapter presents an example of how a conservation 'valuation' approach is mobilised in practice. Insights gained and implications highlighted bear relevance for how similar value-based approaches or economistic methodologies may play out in other public policy fields, such as those under consideration in this volume.

Shaping the calculation of 'biodiversity yield'

BDO policy and best practice guidelines (e.g. BBOP 2009, 2012; DEFRA 2012a, 2012b, 2012c, 2012d, 2013; Gardner et al. 2013) are intended to support biodiversity conservation outcomes by providing methodologies for the technical calculation (i.e. quantification) of biodiversity values. These methodologies aim for the determination of losses and gains at different physical sites and temporal moments in a manner that is commensurable, and that thereby creates the possibility of 'offset' compensation for ecological damage caused by site-specific economic development. BDOs inevitably fall across two separate geographic sites. They propose that quantitative biodiversity gains are reflected as units or credits from a baseline over time at a 'receptor' or offset site, in compensation for biodiversity unit losses at a development site. These are additional to a projected 'counterfactual' scenario in the absence of compensation (Bull et al. 2014). Replicating the wording of a respondent in our research (see below) we use the term 'biodiversity yield' to describe these projected gains in biodiversity values, calculated according to the balance of 'biodiversity units' attained through application of a calculative device.

A variety of metrological approaches exist for calculating and creating equivalence between biodiversity losses and gains at different sites and temporal moments. In application, these are normally linked with use of a standardised reference system for the classification of habitats and/or land cover (Quintero and Mathur 2011, 1122). In England, as further discussed below, the basis for such calculations is the BDO metric developed by DEFRA (2012a), referred to in practice as the Biodiversity Impact Assessment or BIA.

As Habib et al. (2013, 1313–1314) state, '[e]xchanging dissimilar biodiversity elements requires assessment via a generalised metric' and substitution of biodiversity elements through an appropriately fungible currency, or system of credits. A stated aim of BDO is thus the standardisation of state and private sector biodiversity auditing methodologies so as to improve and stabilise approaches considered ad hoc in practice (Gardner et al. 2013, 1254). At the same time, as detailed in this chapter, it is noticeable that the metric as a standardising assessment technique and calculative device in application is being adjusted in creative and diverse ways, such that direct comparisons of offset quality between contexts become difficult. This observation also holds for assessment devices in carbon accounting (see Lohmann 2009; and Lippert 2014, 39). Indicative tensions between ease of compliance for development interests and robustness of conservation gain in terms of measurable biodiversity yield (Maron et al. 2012) again make empirical studies of BDO in application relevant. In particular, such studies can enhance understanding of how these tensions are worked out in practice, given the metrological and policy tools at hand.

The DEFRA BDO metric is adapted from one developed in State of Victoria offsetting programmes in Australia. This quantifies biodiversity values for 'habitat hectares' (Parkes et al. 2003) and is used for calculating and making commensurable losses and gains of biodiversity in development and potential offset sites (DEFRA 2012a; see review in Sullivan 2013). In applying the metric, development sites are first 'mapped and divided into habitat parcels' (DEFRA 2012a, 7) and then classified according to the habitat designations of the Joint Nature Conservation Committee, a public body that advises the UK central and devolved governments on nature conservation.[1] The classification of habitats is made according to three attributes characterising the habitat hectare methodology. These attributes are *spatial area, habitat distinctiveness* and *condition* (DEFRA 2012a). DEFRA provides a gridded matrix for the last two attributes so as to allocate spatial units of mapped habitats to one of three banded scores (Table 7.1). The official BDO scoping report forming the basis of this technical guidance by Treweek et al. (2009, 118) recognised that 'a larger matrix might give a closer fit to reality but would be less straightforward to apply in practice'. Indeed, the BDO metric agreed by DEFRA was itself reduced in complexity from the four habitat condition bands originally proposed (in GHK and eftec 2011, 16) to three in the device proposed by DEFRA (Sullivan 2013, 85). By scoring habitats in this way, the metric quantitatively aligns a score for the qualities of a habitat's ecological distinctiveness with a score for its condition. Through this mechanism, a value for a hectare of biodiversity habitat is generated as a numerical surrogate with a score of between 2 and 18.

Table 7.1 Habitat scoring system for biodiversity offsetting in England

		Biodiversity distinctiveness		
		Low (2)	*Medium (4)*	*High (6)*
Habitat condition	Good (3)	6	12	18
	Moderate (2)	4	8	12
	Poor (1)	2	4	6

Source: DEFRA (2012a, 7).

Biodiversity offsetting in England

Together with the United States, Australia, South Africa and Germany, England is considered to be at the forefront of developing BDO. BDO in England has been enthusiastically endorsed at ministerial level (DEFRA 2013), as well as in a number of recent environmental policy documents and reports (Lawton et al. 2010; DEFRA 2011a, 2013; DCLG 2012). DEFRA set up a BDO pilot that ran for two years from April 2012 to April 2014 and involved six local planning authorities (Devon; Doncaster; Essex; Greater Norwich; Nottinghamshire; and Coventry, Warwickshire and Solihull), as well as various private sector organisations (DEFRA 2011b; for discussion see Hannis and Sullivan 2012; Sullivan 2013; Carver 2015; Apostolopoulou and Adams 2015; Sullivan and Hannis 2015; Ferreira 2017).

Although it had been under discussion prior to 2010 (Treweek et al. 2009), BDO came to fruition under the new UK coalition government which formed in May that year. The Conservative-led coalition government came to power amid a spate of environmental commitments, declaring theirs would be the 'greenest government ever'.[2] The moment was animated by a convergence of factors that produced a wider 'green' political economy discourse shaping the context in which BDO was initially justified (see, for example, Wilshusen and MacDonald 2017). These conditions included the stabilising of a consensus over ecosystem service and biodiversity valuation as the basis for environmental markets to stimulate innovative sources of funding for conservation under a pronounced climate of economic austerity (Comerford et al. 2010; TEEB 2010). In addition to this alignment, a discursive turn towards marrying economic and ecological goals through the 'green economy' dovetailed with a powerful agenda for economic 'recovery' reflected through extensive business de-regulation. One of the hallmarks of this de-regulation was the introduction of a 'one-in two-out rule' which stated that '[w]hen policymakers do need to introduce a new regulation, and where there is a cost to complying with that regulation, they have to remove or modify an existing regulation with double the cost to business'.[3] Under this rule the UK government sought to reduce the cost of regulation to businesses. At the same time as the Department for Communities and Local Government (DCLG) was reforming planning policy through the publication of the National Planning Policy Framework (NPPF) (DCLG 2012), DEFRA announced its intention to trial

BDO with a pilot study in line with this agenda. Among various other modifications, the NPPF included an ideological commitment to a 'presumption in favour of sustainable development' and introduced the concept of a 'no net loss' of biodiversity to planning policy (DCLG 2012) to be managed through application of BDO. Overall then, BDO guidelines were intended to reduce regulatory cost to developers whilst simultaneously conserving biodiversity.

Calculating biodiversity value in practice

We present empirical observations of the actual conducting of biodiversity calculations and valuations for a case study of BDO, as this unfolded in the context of DEFRA's two-year (2012–2014) pilot study. Field research proceeded in three phases and was based on participant observations and semi-structured interviews with 30 key actors through repeat site visits to DEFRA pilot and voluntary offset sites (described in full in Carver and Sullivan 2017; Carver 2017). Textual analysis of 50 semi-structured interview transcripts, secondary public domain and unpublished documents, field-work journals and case study narratives, has been conducted using Dedoose, a cloud-based qualitative data analysis software program. Deductive and inductive coding of text excerpts enabled the emergence of overarching themes and concepts to form the basis of analysis (see, for example, Sullivan and Hannis 2015). Interviewees were grouped into stakeholder categories and subcategories (see Table 7.2). In referring to interview transcripts below, stakeholder categories for the DEFRA pilot sites (as a sample of the six local planning authorities [LPA] that took part in the pilot), are labelled 1–5. The complementary voluntary pilot offsetting site is denoted by a *comp* prefix. Respondent codes are denoted according to the following sequence: stakeholder category, followed by site number, the individual within that site, and date of interview (e.g. LPA2.3 130515 denotes the third individual interviewed within the LPA stakeholder category at pilot site 2, on 13 May 2015).

Below we detail the process of making biodiversity calculations to negotiate biodiversity values in terms of losses, gains (or 'yields'), and prices for

Table 7.2 Categories and subcategories (abbreviations in parentheses) of stakeholders interviewed relative to biodiversity offsetting in England

Category	*Subcategory*
Government:	1. local planning authority (LPA)
	2. Natural England (NE)
Private sector:	3. developer (DEV)
	4. consultant ecologist (CE)
	5. planning consultant (PC)
Civil society:	7. conservation and wildlife NGO (NGO)
	8. local resident (LR)
	9. landowning offset provider (OP)

different nodes of the offset agreement for a particular offset contract – as these were negotiated over 32 months between March 2013 and December 2015. We draw on multiple sources of data acquired through participant observations, semi-structured interviews and analysis of planning documents (described fully in Carver and Sullivan 2017) to document negotiated outcomes for the development site, and to present in detail the metric calculations in the BIA applied to the development and offset sites in this case. As discussed below, the different iterations of the BIA have implications for the scored values arising through its application, independently of the biophysical dimensions of the habitats being assessed.

The development site subject to biodiversity calculations

The case study followed a planning application in England for the delivery of 200 residential properties along with a new sports stadium, and a range of playing fields across 13 ha of mostly agricultural fields (for maps see Figure 1 in Carver and Sullivan 2017). Under the guidance of the DEFRA BDO pilot, and with the assistance of a private offsetting brokerage firm, the development became subject to BDO compensation payments in line with the calculated value of affected biodiversity at the development site. The baseline habitats of the development site consist largely of amenity and improved grasslands, hedgerows, scattered tall ruderal vegetation and four ponds, one with great crested newts (*Triturus cristatus)* which are protected under the UK's Conservation of Habitats and Species Regulations 2010. The site forms the south-western fringe of a small medieval market town. It is bordered to the north by an industrial and residential development, and to the west and east by roads with open countryside beyond. The planning application was submitted in March 2013 by the existing owners, the local football and bowls club, in conjunction with a large residential developer who would oversee the bulk of the planning process and through whom the residential properties would be built, marketed and sold. Under new requirements shaped by the DEFRA BDO pilot study, the local planning authority asked the developer to apply the DEFRA metric, in the form of a BIA, to guide the biodiversity mitigation and compensation measures required to offset the development.

The BIA in practice

In the case explored here, the developers' consultant ecologists completed primary habitat surveys in addition to using existing records derived from earlier surveys of the site as the basis for an Ecological Impact Assessment (EIA) report for the planning application. Subsequently, as a desk-based exercise and to establish the site's biodiversity unit baseline and mitigation values in a format required for BDO under the DEFRA pilot, the contents of the EIA were translated into the BIA by a LPA ecologist and the offset broker assisting the LPA with the contract. As the calculative device, the

BIA computed the ecological condition and proposed mitigation works at the development site. This process entailed re-codifying the existing habitats into an Excel spreadsheet supplied by the County Council planning authority ecologist acting as the DEFRA pilot site lead. The BIA spreadsheet operationalises the DEFRA BDO metric to arrive at biodiversity unit scores for each bounded habitat type on the development site.

In order to arrive at these scores the development site was first categorised by the LPA ecologist into habitat types entered on separate rows of the BIA spreadsheet, each with a code, description, size in hectares, and numerical scores for habitat distinctiveness and condition (Table 7.3 shows the first BIA draft on 27 June 2013, prior to re-negotiation). The spreadsheet enables three sets of calculation. The first (Table 7.3, rows 15–28) generates a 'Habitat Impact Score' (HIS) for the total scored habitat on site prior to the development, in this case 46.68 biodiversity units (cell O53).

The second step shown in Table 7.3 (rows 56–89) calculated the 'Habitat Mitigation Score' (HMS) as the total number of biodiversity units that will be restored or created on-site so as to 'mitigate' or minimise projected biodiversity losses, in this case 16.78 units (Cell O89). In this case, a notable feature of the HMS is that almost 40 per cent of the on-site mitigation work was to be delivered through creation of a number of football pitches of varying sizes. The presence of these football pitches was calculated as contributing significant amounts of on-site biodiversity unit value and the largest habitat spatial value of all mitigation activities. This mitigation value was to be achieved by attributing biodiversity value to the amenity grassland of the pitches themselves (cell O63), through permitting the perimeter surface area of their proposed grassy margins to grow un-mowed and develop into semi-improved grassland (cell O70). Although receiving low habitat scores for both distinctiveness (2) and habitat condition (1), the size of the football pitches amenity grassland meant that in five years this aggregate *spatial* area was calculated to contribute 4.07 units. After 15 years the grassy margins would contribute a further 2.29 units. The football pitches therefore would make up 6.36 units of biodiversity value out of 16.78 units of total on-site mitigation and habitat creation (Figure 7.1).

In the third step, and through comparing the scores from steps 1 and 2 above, the residual biodiversity 'net loss' was calculated. This value indicates how many units would be required for mitigation purposes via an offset where additional biodiversity conservation value is purchased. This 'net' loss or gain value is the Habitat Biodiversity Impact Score (HBIS), and is generated through the equation:

$$\text{Habitat Mitigation Score (HMS)} - \text{Habitat Impact Score (HIS)} \\ = \text{Habitat Biodiversity Impact Score (HBIS)}$$

In this case the HBIS is calculated as an overall loss of 31.90 biodiversity units (Table 7.3 cell O91). It is this calculated biodiversity value that requires an off-site offset to satisfy planning requirements for mitigation of development impacts on biodiversity.

Table 7.3 Spreadsheet (version17.4, draft 1) used in biodiversity impact assessments to calculate the residual losses of biodiversity from development impacts[a]

A[a]	*B*	*C*	*D*	*E*	*F*	*G*	*H*	*I*	*J*	*K*	*L*	*M*	*N*	*O*
11														
12		*Existing habitats on site*			*Habitat distinctiveness*		*Habitat condition*		*Habitats to be retained with no change within development*		*Habitats to be retained and restored within development*		*Habitats to be lost within development*	
13	*T.note*[b]	*habitat code*	*phase 1 habitat description*	*habitat area (ha)*	*Distinctiveness*	*score*	*condition*	*score*	*area (ha)*	*existing value*	*area (ha)*	*existing value*	*area (ha)*	*existing value*
14			direct impacts and retained habitats			A		B	C	A × B × C = D	E	A × B × E = F	G	A × B × G = H
15	F1	B4	grassland: improved grassland	1.78	low	2	moderate	2					1.78	7.12
16	P1	G1	wetland: standing water	0.01	high	6	good	3					0.01	0.14
17	P1	B4	grassland: improved grassland	3.10	low	2	moderate	2					3.10	12.40
18	F2	C31	grassland: amenity grassland	0.18	low	2	moderate	2					0.18	0.70
19	Bowling green	J12	grassland: amenity grassland	0.12	low	2	poor	1					0.12	0.24
20	F6	J12	grassland: amenity grassland	0.97	low	2	poor	1					0.97	1.94
21	West of football	B4	grassland: improved	0.08	low	2	good	3					0.08	0.48

			environment: buildings/ hardstanding	0.32	none	0	poor	1					0.32	0.00	
26	F3	B4	grassland: improved grassland	4.88	low	2	moderate	2					4.88	19.52	
27	Slurry pit	G1	wetland: standing water	0.01	high	6	poor	1					0.01	0.05	
28	Part of F4 and F5	B4	grassland: improved grassland	1.52	low	2	moderate	2					1.52	6.08	
45	Total			12.96				Total	0.00	0.00	0.00	0.00	12.96	48.68	
46														ΣD+ΣF+ΣH	
47												site habitat biodiversity value		46.68	J
48	Indirect impacts, including offsite habitats			K					value of loss from indirect impacts[c]						
49	Before or after impact								K × A × B = Li, Lii	Li- Lii					
50	Before														
51	After														
52	Total			0.00					M	0.00				HIS = J + M	
53													habitat impact score (HIS)	46.68	
54	Caution: Destruction of habitats of high distinctiveness, e.g. lowland meadow or ancient woodland, may be against local policy. Has the mitigation hierarchy been followed, can impact to these habitats be avoided? Any unavoidable loss of habitats of high distinctiveness must be replaced like for like.														
55															

(*Continued*)

A[a]	*B*	*C*	*D*	*E*	*F*	*G*	*H*	*I*	*J*	*K*	*L*	*M*	*N*	*O*
56		*Proposed habitats on site (onsite mitigation)*			*Target habitat distinctiveness*		*Target habitat condition*			*Time till target condition*		*Difficulty of creation or restoration*		*Habitat biodiversity value*
57			*phase 1 habitat description*	*Habitat area (ha)*	*distinctiveness*	*score*	*condition*	*score*		*time*	*score*	*difficulty*	*score*	
58			habitat creation	N	O		P				Q		R	(N × O × P)/Q/R
59	F1 & F6	n/a	Built environment: buildings orhardstanding	2.27	none	0	poor	1		5 years	1.2	low	1	0.00
60	F1 & F6	n/a	built environment garden (lawn and planting)	0.76	low	2	poor	1		5 years	1.2	low	1	1.27
60	F1 & F6	n/a	built environment garden (lawn and planting)	0.76	low	2	poor	1		5 years	1.2	low	1	1.27
61	F2, F3, F4	n/a	built environment: buildings or hardstanding	4.02	none	0	poor	1		5 years	1.2	low	1	0.00
62	F2, F3, F4	n/a	built environment: gardens (lawn and planting)	1.34	low	2	poor	1		5 years	1.2	low	1	2.23
63		J12	grassland: amenity grassland	2.44	low	2	poor	1		5 years	1.2	low	1	4.07
64	Area 4	B22	grassland: semi-improved neutral grassland	0.76	medium	4	good	3		15 years	1.7	medium	1.5	3.58
65	Area 3	G1	wetland: standing water	0.13	high	6	good	3		15 years	1.7	medium	1.5	0.92

			semi-improved neutral grassland											
67		G1	wetland: standing water	0.05	high	6	good	3		15 years	1.7	medium	1.5	0.35
68	Landscape	A112	woodland: broadleaved plantation	0.16	medium	4	good	3		32+ years	3	medium	1.5	0.43
69	Area 2	B22	grassland: semi-improved neutral grassland	0.19	medium	4	good	3		15 years	1.7	medium	1.5	0.89
70		B22	grassland: semi-improved neutral grassland	0.73	medium	4	moderate	2		15 years	1.7	medium	1.5	2.29
71	Total			13.01	ERROR: total area of habitats created must equal total area of habitats lost									
72	Habitat restoration								existing value S (=F)					((N × O × P) – S)/Q/R
88			Total							trading down correction value				0.00
89										habitat mitigation score (HMS)				16.78
90														HBIS–HMS = HIS
91										habitat biodiversity impact score				–31.90
92										percentage of biodiversity impact loss				65.53

Notes

[a] Numbers across rows 15–28 are multiplied to produce the habitat biodiversity value for each coded area subject to development. Numbers are rounded to 2 decimal places and, as such, if calculated manually here may produce different results. The sum of the values in column O for rows 15–28 (shown in cell O53) is the biodiversity baseline of the habitat impact score (HIS). In rows 59–70 numbers across rows are multiplied to produce the habitat-mitigation score (HMS) (cell O89), which is the value of habitats that will be restored or created onsite so as to mitigate or minimise projected biodiversity losses.

[b] T note refers to the ‘target’ area as habitat parcels subject to impacts coded on the development maps.

[c] Left empty on original BIA.

	Development Site	Offset Site
Full size football pitches (number)		 (Area of land equivalent in size)
Size (ha)	2.97	0.66
Habitat type (qualitative category)	Amenity grassland	Species-rich semi-improved grassland developed into a lowland meadow
Biodiversity units	4.07	5.6

Figure 7.1 One football field with many 'fields' of value

The offset site

Established conservation NGOs and Wildlife Trusts are often preferred partners in delivering compensatory conservation to mitigate the development impacts. These organisations are typically perceived to possess the right 'experience and expertise to ensure delivery' (LPA2.1 020714), as well as being familiar and experienced with this type of contracted management for habitat creation. In this case, the offset site identified by the council for the supply of scored offset units comparable to the HBIS score above was a 5 ha grassland meadow 5 km north-east of the development acquired by a local wildlife and conservation NGO in 2013. The site consists of a meadow of species-rich, semi-improved grassland in close proximity to a local Site of Special Scientific Interest (SSSI). The site currently supports five species of orchid including the largest population of greater butterfly orchid (*Plantanthera chlorantha*) in the county. It also contains four of the county's six rare farmland butterflies, three of which (the grizzled skipper [*Pyrgus malvae*], the dingy skipper [*Erynnis tages*] and the white-letter hairstreak [*Satyrium w-album*]) are nationally designated as biodiversity priority species under the UK Biodiversity Action Plan 2007.

Despite demonstrating excellent ecological enhancement potential, limited funding meant that little management input was envisaged by the NGO at the time of acquisition of the site (NGO5.1 241114). A BIA was applied for the offset site to calculate its baseline and projected biodiversity values in a parallel way to the development site above. This BIA is not reproduced here but indicates that the NGO had intended to bring the grassland into a moderate condition and 'return' the site to a fully favourable species-rich lowland meadow condition of national importance with the offset payments from the developer. These improvements would thereby demonstrate conservation additionality against a projected counterfactual scenario (NGO5.1 Offset Site Draft Management Plan November 2013). In this case it was hoped that, with appropriate interventions, two other rare farmland butterflies, also designated nationally as biodiversity priority species, would establish colonies on the site (NGO 5.1 Offset Site Draft Management Plan November 2013).

In order to achieve this conservation additionality, the predicted biodiversity gain or 'yield' was quantified and costed according to a management plan for the works to be supplied by the offset provider. The BDO Draft Management Plan, written by the NGO conservation officer, estimates costs based on the combined workforce of NGO volunteers, conservation staff and external contractors, materials and capital or lease payments for the land itself (NGO5.2 280115). In this case, the predicted budget for 30 years' management at the offset site put the total figure of habitat work costs at £204,076, of which £98,030 was planned to come from the conservation NGO budget, and £106,046 from the biodiversity offset payment made by the developers (NGO5.1 Offset Site Draft Management Plan November 2013). The developer would pay a further 20 per cent in broker fees (OB5.1 241114), plus legal fees for arranging the contracts between parties (NGO5.2 280115).

Negotiating biodiversity calculations into 'something we can live with'

Successive calculations traced over time for scored HMS, HIS and HBIS values show that the stakeholders negotiated so as to arrive at the costs of compensation. These negotiations frequently involved changing the numerical scores entered into the first BIA for the development site. Over the course of the planning process in this case, the calculated baseline value of the development site was reduced by almost 48 per cent from 48.68 units in BIA draft 1 (Table 7.3) to 25.52 units in draft 2. This downward recalculation occurred through making category changes to the condition of existing habitats, thereby adjusting the numerical unit outcomes of assigned habitat values.

Table 7.4 shows the changes made to the condition values of the baseline habitats on-site and the dramatic effects on financial compensation requirements, identified through comparisons made from different drafts of the BIA Excel sheets. These changes included lowering the perceived condition of 4.88 ha of improved grassland on the development site from a 'moderate' to 'poor' condition, such that the BIA F3 baseline value of 19.52 on Table 7.3

Table 7.4 Changes in calculated biodiversity baselines between Biodiversity Impact Assessments drafts 1 and 2

BIA Habitat Area Code	*Habitat description*	*Area (ha)*	*Distinctiveness*		*Condition*		*Original unit value*	*New unit value*	*Percentage reduction in unit value*
			Category change	*Score change*	*Category change*	*Score change*			
F1	Improved grassland	1.78	None	None	Moderate to poor	2 to 1	7.12	3.56	50%
P1	Wetland-standing water	0.01	None	None	Good to moderate	3 to 2	0.14	0.01	93%
F2	Improved grassland	3.10	None	None	Moderate to poor	2 to 1	12.40	6.20	50%
F2	Tall Ruderal	0.18	None	None	Moderate to poor	2 to 1	0.70	0.35	50%
The Bowling green	Amenity grassland	0.12	None	None	None	None	0.24	0.24	0%
F6	Amenity grassland	0.97	None	None	None	None	1.94	1.94	0%
West of football grid	Improved grassland	0.08	None	None	Good to moderate	3 to 2	0.48	0.32	33%
F3	Improved grassland	4.88	None	None	Moderate to poor	2 to 1	19.52	9.76	50%
Slurry pit	Standing water	0.01	High to low	6 to 2	None	None	0.05	0.02	60%
Part of F4 and F5	Improved grassland	1.52	None	None	Moderate to poor	2 to 1	6.08	3.04	50%
Total							48.68	25.52	48%

(cell O26) became 9.76 units. Due to its size, this constituted a 50 per cent reduction of baseline biodiversity value for this habitat area. The same process was applied to improved grasslands in different compartments in habitat area codes F1 and F2. Modifications regarding anomalous local level category values (e.g. concerning what value scattered trees should have) were also made in anticipation of the numerical and financial outcomes that would arise through the multiplication or division effects of these modifications on biodiversity loss (HIS) or on-site mitigation (HMS). In other words, the ability of these modifications to either enlarge or shrink the final compensation costs of development planning applications was built into the numerical adjustments that followed. Other iterations to the BIA over the course of the pilot study included adding category values to the BDO metric (Table 7.1) with the odd numbers 1, 3 and 5 for local habitat types that have greater regional than national distinctiveness and rarity (NGO5.1 241114), as well as extensive formatting changes and editions to make the calculator more 'user-friendly' and manageable (OB5.2 050114).

One explanation for these modifications is that local planning authorities have a relatively limited ability to determine the exact condition of the habitats under discussion. Only a handful of habitats will have been visited and verified by a county ecologist (LPA5.2 241114), due to the widespread shortage of ecological expertise within local governments more generally (NGO 270116, NGOCompl 2901016). Often the ecological data is assembled and cross-referenced remotely against a combination of records supplied in the Ecological Assessment Report in addition to biological and historical archives held on file at the council offices at the district council's Biological Records Centre. Indeed, in this instance the good quality and extensive scope of the ecological data held by the County Council for reference and the size of its own ecological team are considered to be anomalous within England (LPA5.2 241114).

At the same time, underlying the overall downwards recalculation of the development site's baseline value from 48.68 to 25.52 units was a view that the first BIA calculations would create too large a compensation package, thereby threatening the financial viability of the residential development (DEV5.1 060315). The initial calculation for the biodiversity offset compensation package was £300,000. During the course of an hour-and-a-half meeting between the developer and the local planning authority, and at the instigation of the development firm, the baseline habitat condition assessments for many areas on-site were adjusted downwards to produce the new figure of 25.52 biodiversity units in the second draft (Table 7.4). Described as 'something we could live with' (DEV5.1 060315), the ensuing revised cost for the final compensation package was £120,000.

Discussion: working towards the right balance of values

This chapter provides an in-depth case history of the assessments, calculations and negotiations of biodiversity and financial values for a BDO compensation contract *in situ*. Our case study illuminates several problems for biodiversity conservation predicted in the theoretical literature on BDO. In conclusion we highlight three points in particular, and focus on their broader implications for biodiversity conservation through the implementation of BDO.

First, evidence from repeat site visits to this and other pilot offsetting cases in England over 24 months shows that the business of making the DEFRA metric applicable to real-life planning cases has proven to be a process of constant iteration, trial and error. The BDO case history documented above illustrates in particular how metrics for determining biodiversity values at development and receptor sites are being redesigned in application and are generating numerical values that are then further negotiated and adjusted. The model of BDO and the devices in service to this are envisaged to produce a neutral calculative framework and a technical means for making impartial and objective decisions based around value. As our case study shows, however, the new metrics associated with BDO are being used in creative ways by different actors negotiating competing requirements. This creativity may be appropriate in response to real-world complexity. Nonetheless, it is also in tension with stated aims in BDO policy design for standardisation and comparability. In this case, the nexus of competing development, conservation and LPA interests meant that biodiversity values calculated through application of the metric were adjusted downwards so as to facilitate a compensation package that was cheaper for developers. The firm had been involved in another negotiation in connection with a different site, that also involved a reduction of biodiversity offset compensation value, from £300,000 to £90,000 in an even shorter 30-minute meeting (DEV5.1 06031; also see case documented in Sullivan 2013). Thus, as predicted in theory (Walker et al. 2009; Hannis and Sullivan 2012), these case observations seem consistent with concerns that an emphasis on market values for biodiversity conservation and compensation will encourage developers, as purchasers of impact compensation, to push prices downwards to lower their costs. In doing so, both the quality and quantity of conservation 'yield' through BDO may also be reduced. The downward adjustment to biodiversity impact values illustrates that valuation algorithms in service to decision-making are only as good as their numerical and category inputs, shown, by this case study, to be subject to economic and political factors.

Second, the case history provides empirical material illustrating the working in practice of commensuration processes making different habitats equivalent to each other through application of BDO metrics (Tables 7.1 and 7.3). Numerical signifiers form proxies for qualitatively different ecological assemblages, calculated with the aid of the DEFRA biodiversity metric as represented by the BIA Excel spreadsheet calculations presented above. Sometimes these commensuration processes generate counterintuitive outcomes. In this case it is unclear how exactly a sports pitch can really be said to be maintaining 6.36 units of biodiversity value at the development site that is equivalent to 6.36 units of high quality grassland habitat supporting a range of Biodiversity Action Plan species at the offset site. The proposed mitigation value provided through sports pitches, measured as habitats of low distinctiveness and poor condition and achieved through acting as the largest 'habitat type' within the development, illustrates the ability of numerical abstractions to commensurate distinct biota of very different qualities (Sullivan 2013; Carver 2015). The design principles of English BDO mean that large habitat areas of low biodiversity value can be treated as equivalent to small habitat areas of high biodiversity value (see Figure 7.1). In this case, the 'packaging' of football pitches as 'habitats' for on-site mitigation again reduced the financial compensation value in the final offset calculations, by decreasing offset unit requirements by an equivalent number of biodiversity units.

Finally, elements of both points above suggest that the pressure of creating value for money in compensation strategies for conservation may be pushing BDO in directions that favour the growth of market values and exchanges for offset units, but that may work against the robust generation of conservation value (Hannis and Sullivan 2012; Dauguet 2015). Additional observations in our case study confirm this tendency. For example, although conservation NGOs are considered to be convenient offset providers or 'low hanging fruit' by district councils (LPA5.1 010215), as in the case presented here, councils also recognise that in order to achieve market liquidity in offset units, private landowners will have to play a significant role. Indeed, a barrier to the development of BDO in England has been a shortage of 'supply side' offset sites from which to purchase biodiversity offset credits, as confirmed in DEFRA's pilot evaluation report (Baker et al. 2014). The local authority in the above case has thus engaged partnerships with organisations that can assist in stimulating and brokering a greater supply of offset credits and receptor sites through private landowners. High transaction costs from site identification, preparation and legal fees are fostering economies of scale by bringing offset provision together with the economically astute farm management strategies of large commercial landowners (OB5.2 020315; and predicted in Sullivan and Hannis 2015). This is expected to result in a better supply of offset credits from newly formed 'habitat banks'.

Emerging offset brokerage firms hope that a high supply of offset credits will improve market competition, resulting in cheaper prices for developers seeking compensation. This combination of commercial outlook, farming experience and land management makes the corporate broker now partnering with the county council confident that with this approach they can produce a 'good biodiversity yield per hectare' across multiple sites (OB5.2 020315). In doing so, BDO is becoming further aligned with commercial agricultural productivity agendas that emphasise efficiencies and scale of production through concepts of agricultural 'yield'.

The case presented here is of a BDO to be provided directly by a conservation NGO to a UK local planning authority. Despite months of preparation and considerable staff costs, the developer eventually rejected the proposed offset site in favour of developing an offset arrangement with the farmer issuing the land for development in the first place (OB5.1 140116). The calculations and negotiations presented here are nonetheless valuable as a detailed example of how the DEFRA metric is being applied in practice to commensurate biodiversity assessments between different sites. In tracking, documenting and analysing the calculations and negotiations in this and other cases, we observe that although BDO is constituted by technical and apolitical practices to calculate equivalence and commensurability between sites of biodiversity damage and conservation investment, the way these are enacted in practice is subject to a proliferation of methods, techniques and valuation criteria that are balanced to meet conflicting user requirements.

As such, instead of confining the decision-making process to a neutral calculative and technical framework, the technical process itself opens new possibilities through which stakeholders negotiate, and sometimes struggle over, specific outcomes (Coralie et al. 2015; Sullivan and Hannis 2015).

These spaces and the political conditions in which they are produced are obscured within the black boxes of idealised calculative models and final-draft 'valuations'. As Walker et al. (2009, 149) identify, the concern is that 'biodiversity protection interests will fail to counter motivations for officials to resist and relax safeguards to facilitate exchanges and resource development at cost to biodiversity'. Application of scoring practices that create numerical values may help stimulate greater compensation payments for biodiversity loss in the English planning system and elsewhere, thereby supporting greater biodiversity outcomes (however, see Carver 2015). Obscured within these technical practices in the application of a calculative device, however, and as predicted by Salzman and Ruhl (2000), are additional value judgments and struggles over arriving at the perceived 'right' numerical values. Our empirical observations show that these values must be able to straddle competing demands such that they are economically palatable, politically pragmatic as well as ecologically coherent. Whether or not biodiversity 'yields' are achieved through these negotiations will depend on the bargaining powers of those involved, beyond the application of standardised practices to calculate and commensurate biodiversity losses and gains.

There is a broader postscript to the case history shared here. At the close of the DEFRA pilot study in 2014 it was expected that DEFRA would release a policy statement following the publication of its evaluation report (Baker et al. 2014), and yet no such announcement materialised. Amongst other things, the evaluation of the pilot proposed that a system of BDO in England would likely increase the overall costs of compensation to developers in the short and long term (Baker et al. 2014). As such, BDO would neither align with the promises to the construction industry made in the Conservative manifesto, nor meet requirements from HM Treasury for de-regulation and the 'one-in two-out rule' (stepped up to one-in three-out the following year) as described above. Our case study illustrates the ways in which a 'value for money' requirement appears as an influence shaping the application of the calculative practices at the level of individual BDO contracts. In the end, however, it appears that BDO as a regulatory approach could not prove its own 'value for money' within a political climate of de-regulation and economic growth. From 2014 onwards, DEFRA quietly let the spotlight fade from the BDO approach as a legislated policy requirement for the management of development impacts in England. BDO in a voluntary capacity, however, continues to unfold within local government and corporate contexts. As a calculative approach with a focus on measurability BDO thus remains as the only mechanism now available to meet policy requirements in the NPPF for demonstrating measurable 'no net loss' of biodiversity 'value'.

Notes

1 See http://jncc.defra.gov.uk
2 See transcript of 2010 speech by then Prime Minister David Cameron at https://www.gov.uk/government/speeches/pms-speech-at-decc (Accessed 2 August 2017).
3 See https://www.gov.uk/government/publications/2010-to-2015-government-policy-business-regulation/2010-to-2015-government-policy-business-regulation#appendix-4-operating-a-one-in-two-out-rule-for-business-regulation (Accessed 2 August 2017).

References

Apostolopoulou A. and Adams A. (2015) "Biodiversity offsetting and conservation: reframing nature to save it" *Oryx*, 51(1): 1–9.

Baker J., Sheate W.R., Bennett T., Payne D., Tucker G., White O. and Forrest S. (2014) *WC1051: Evaluation of the biodiversity offsetting pilot program. Final report volume 1* Department for Environment, Food and Rural Affairs, London.

BBOP (Business and Biodiversity Offsets Programme) (2009) *Biodiversity offset design handbook* BBOP, Washington, DC.

BBOP (Business and Biodiversity Offsets Programme) (2012) *Standard on biodiversity offsets* BBOP, Washington, DC.

Benabou S. (2014) "Making up for lost nature? A critical review of the international development of voluntary biodiversity offsets" *Environment and Society: Advances in Research*, 5: 102–123.

Bull J.W., Gordon A., Law E.A., Suttle K.B. and Milner-Gulland J. (2014) "Importance of baseline specification in evaluating conservation interventions and achieving no net loss of biodiversity" *Conservation Biology*, 28: 799–809.

Çalışkan K. and Callon M. (2009) "Economization, part 1: shifting attention from the economy towards processes of economization" *Economy and Society*, 38: 369–398.

Çalışkan K. and Callon M. (2010) "Economization, part 2: a research programme for the study of markets" *Economy and Society*, 39: 1–32.

Callon M. (1998) "An essay on framing and overflowing: economic externalities revisited by sociology" in Callon M. (ed.) *The laws of markets* Blackwell, Oxford, 244–269.

Callon M. and Muniesa F. (2005) "Peripheral vision: economic markets as calculative collective devices" *Organization Studies*, 26: 1229–1250.

Carroll N., Fox J. and Bayon R. (eds) (2008) *Conservation and biodiversity banking: a guide to setting up and running biodiversity credit trading systems* Earthscan, London.

Carver L. (2015) *Measuring the value of what? An ethnographic account of the transformation of 'Nature' under the DEFRA biodiversity offsetting metric* LCSV Working Paper 11, http://thestudyofvalue.org/wp-content/uploads/2015/02/WP11-Carver-2015-Measuring-whatvalue.pdf (Accessed 4 September 2017).

Carver L. (2017) Assembling the value of nature: a performative analysis of English biodiversity offsetting and the DEFRA pilot study. Unpublished PhD thesis, Department for Geography, Environment and Development, Birkbeck, University of London.

Carver L. and Sullivan S. (2017) "How economic contexts shape calculations of 'yield' in biodiversity offsetting" *Conservation Biology*, doi:10.1111/cobi.12917

Comerford E., Molley D. and Morling P. (2010) *Financing nature in an age of austerity* Royal Society for the Protection of Birds (RSPB) [online] https://www.rspb.org.uk/Images/Financingnature_tcm9-262166.pdf (Accessed 4 September 2017).

Coralie C., Guillaume O. and Claude N. (2015) "Tracking the origins and development of biodiversity offsetting in academic research and its implications for conservation: a review" *Biological Conservation*, 192: 492–503.

Dauguet B. (2015) "Biodiversity offsetting as a commodification process: a French case study as a concrete example" *Biological Conservation*, 192: 533–540.

DCLG (Department for Communities and Local Government) (2012) National planning policy framework (NPPF), https://www.gov.uk/government/uploads/system/uploads/attachment_data/file/6077/2116950.pdf (Accessed 4 September 2017).

DEFRA (Department for Environment, Food and Rural Affairs) (2011a) *The natural choice: securing the value of nature* UK Government, London, https://www.gov.uk/government/uploads/system/uploads/attachment_data/file/228842/8082.pdf (Accessed 4 September 2017).

DEFRA (Department for Environment, Food and Rural Affairs) (2011b) *Biodiversity offsetting. Testing biodiversity offsetting* UK Government, London.

DEFRA (Department for Environment, Food and Rural Affairs) (2012a) *Biodiversity offsetting pilots technical paper: the metric for the biodiversity offsetting pilot in England* UK Government, London, https://www.gov.uk/government/uploads/system/uploads/attachment_data/file/69531/pb13745-bio-technical-paper.pdf (Accessed 4 September 2017).

DEFRA (Department for Environment, Food and Rural Affairs) (2012b) *Biodiversity offsetting pilots: information note for local authorities* UK Government, London, https://www.gov.uk/government/uploads/system/uploads/attachment_data/file/69529/pb13744-bio-local-authority-info-note.pdf (Accessed 4 September 2017).

DEFRA (Department for Environment, Food and Rural Affairs) (2012c) *Biodiversity offsetting pilots: guidance for developers* UK Government, London, https://www.gov.uk/government/uploads/system/uploads/attachment_data/file/69528/pb13743-bio-guide-developers.pdf (Accessed 4 September 2017).

DEFRA (Department for Environment, Food and Rural Affairs) (2012d) *Biodiversity offsetting pilots: guidance for offset providers* UK Government, London, https://www.gov.uk/government/uploads/system/uploads/attachment_data/file/69530/pb13742-bio-guide-offset-providers.pdf (Accessed 4 September 2017).

DEFRA (Department for Environment, Food and Rural Affairs) (2013) *Biodiversity offsetting in England Green Paper* UK Government, London, https://consult.defra.gov.uk/biodiversity/biodiversity_offsetting/supporting_documents/20130903 Biodiversity%20offsetting%20green%20paper.pdf (Accessed 4 September 2017).

Espeland W.M. and Stevens M.L. (1998) "Commensuration as a social process" *Annual Review of Sociology*, 24: 313–343.

Ferreira C. 2017 "The contested instruments of a new governance regime: accounting for nature and building markets for biodiversity offsets" *Accounting, Auditing and Accountability Journal,* doi.org/10.1108/AAAJ-12-2015-2336

Fox J. and Nino-Murcia A. (2005) "Status of species conservation banking in the United States" *Conservation Biology*, 19: 996–1007.

Gardner T.A., Von Hase A., Brownlie S., Ekstrom J.M.M., Pilgrim J.D., Savy C.E., et al. (2013) "Biodiversity offsets and the challenge of achieving no net loss" *Conservation Biology,* 27: 1254–1264.

GHK and eftec (2011) *Costing potential actions to offset the impact of development on biodiversity. Final report for DEFRA* GHK Consulting Ltd, London.

Habib T.J., Farr D.R., Schneider R.R. and Boutin S. (2013) "Economic and ecological outcomes of flexible biodiversity offset schemes" *Conservation Biology*, 27: 1313–1323.

Hannis M. and Sullivan S. (2012) *Offsetting nature? Habitat banking and biodiversity offsets in the English land use planning system* Green House, Weymouth, Dorset.

Helgesson C.F. and Muniesa F. (2013) "For what it's worth: an introduction to valuation studies" *Valuation Studies*, 1: 1–10.

Helm D. (2014) "Taking natural capital seriously" *Oxford Review of Economic Policy*, 30: 109–125.

Helm D. (2015) *Natural capital: valuing the planet* Yale University Press, London.

Lawton J., Brotherton P.N.M., Brown V.K., Elphick C., Fitter A.H., Forshaw J., et al. (2010) *Making space for nature: a review of England's wildlife sites and ecological network*. Consultation presented to Department for Environment, Food and Rural Affairs (DEFRA).

Lippert I. (2014) "Studying reconfigurations of discourse: tracing the stability and materiality of sustainability/carbon" *Journal for Discourse Studies*, 1: 32–54.

Lohmann L. (2009) "Toward a different debate in environmental accounting: the cases of carbon and cost–benefit" *Account, Organisations and Society*, 34: 499–534.

Maron M., Hobbs R.J., Moilanen A., Matthews J.W., Christie K., Gardner T.A., et al. (2012) "Faustian bargains? Restoration realities in the context of biodiversity offset policies" *Biological Conservation*, 155: 141–148.

Muniesa F., Millo Y. and Callon M. (2007) "An introduction to market devices" *Sociological Review*, 55(2 suppl): 1–12.

Parkes D., Newell G. and Cheal D. (2003) "Assessing the quality of native vegetation: the 'habitat hectares' approach" *Ecological Management & Restoration*, 4: S29–S34.

Pawliczek J. and Sullivan S. (2011) "Conservation and concealment in Species Banking.com, USA: an analysis of neoliberal performance in the species offsetting industry" *Environmental Conservation*, 38: 435–444.

Quintero J.D. and Mathur A. (2011) "Biodiversity offsets and infrastructure" *Conservation Biology*, 25: 1121–1123.

Robertson M. and Hayden N. (2008) "Evaluation of a market in wetland credits: entrepreneurial wetland banking in Chicago" *Conservation Biology*, 22: 636–646.

Salzman J. and Ruhl J.B. (2000) "Currencies and commodification of environmental law" *Stanford Law Review*, 53: 607–687.

Sullivan S. (2013) "After the green rush? Biodiversity offsets, uranium power and the 'calculus of casualties' in greening growth" *Human Geography*, 6: 80–101.

Sullivan S. (2017) "Noting some effects of fabricating 'nature' as 'natural capital'" *Ecological Citizen*, 1: 65–73.

Sullivan S. (forthcoming) "Making nature investable: from legibility to leverageability in the making of natural capital" *Science and Technology Studies*.

Sullivan S. and Hannis M. (2015) "Nets and frames, losses and gains: value struggles in engagements with biodiversity offsetting policy in England" *Ecosystem Services*, 15: 162–173.

TEEB (2010) *The Economics of Ecosystems and Biodiversity Ecological and Economic Foundations* Kumar P. (ed.) Earthscan, London.

Treweek J., ten Kate K., Butcher B., Venn O., Garland L., Wells M., et al. (2009) *Scoping study for the design and use of biodiversity offsetting in an English context*. Final Report to DEFRA. Contract NE 0801 https://www.cbd.int/financial/offsets/unitedkingdom-scoping.pdf (Accessed 6 December 2017).

Walker S., Brower A.L.B., Stephens R.T.T. and Lee W.G. (2009) "Why bartering biodiversity fails" *Conservation Letters*, 2: 149–157.

Wilshusen P.R. and MacDonald K.I. (2017) "Fields of green: corporate sustainability and the production of economistic environmental governance" *Environment and Planning A*, 49: 1824–1845.

Part 3

Environment

8 A crash in value

Explaining the decline of the Clean Development Mechanism

Robert Watt

Introduction

Carbon offset markets, which involve the trade of credits in 'emission reductions' typically produced in developing countries, have declined. This chapter offers an explanation based on a number of interconnected meanings of value. The UN Kyoto Protocol established the Clean Development Mechanism (CDM) and Joint Implementation (JI) as 'flexibility mechanisms' that enabled carbon offsetting as part of the international policy response to climate change. The CDM developed into a booming market where investors were making significant returns on 'low carbon' investments. However, at present, the market has crashed with CDM offsets trading at very low exchange values.

The chapter argues that a primary driver for the decline in offset markets is the growing perception that offsets have limited use value as a result of their failure to provide the intended quantified carbon emissions reductions. Most of the offset credits generated under the CDM have been sold to the world's largest carbon market, the European Union Emissions Trading System (EU ETS). However, the demand for carbon credits in the presently over-supplied EU ETS has become strictly limited due to a growing realisation that carbon offsets lack use value.

This failure is related to the complications of the valuation processes involved in creating an offset commodity. These include challenges of measuring emissions reductions through the postulation of hypothetical 'baselines', and problems of attribution of emissions reductions to the offset mechanism. Both involve navigating the complex concept of 'additionality'. The commercial interests and power of actors involved in the valuation process, such as project developers and private sector auditors, encourage the registration of significant amounts of offsets which lack environmental integrity.

Policy-makers and other users of carbon credits evaluate offsetting in diverse ways, which can influence public evaluations of offsets. While some policy-makers are concerned about offsets' lack of environmental integrity, others either do not recognise this at all, or recognise it, but do not perceive it as a problem. Offsets can consequently be framed discursively as valuable, even when they are not, giving rise to an economy of false appearances. Understanding the demand for offsets therefore requires investigation of the normative evaluations of the carbon market among users, in recognition that there is political contestation involved in the construction of those normative discourses.

Before discussing these points in more detail, three denotations of value are briefly considered: exchange value, use value, and valuation. Each roughly corresponds to questions of money, matters of worth, and processes and activities of determining value, respectively. These denotations are applied throughout the discussion.

Denotations of value: exchange, use and valuation

The word 'value' has multiple uses and referents (see Chapter 2, this volume). *Exchange value* deals with issues of price and monetary value (Miller 2008, 1123). If for example you purchase a carbon credit for €5 and either sell it for €6, or swap it for a shiny bracelet, you are dealing with a matter of exchange value. Preoccupation with exchange value – or its close cousins shareholder or monetary value – is a defining feature of capitalism, whereby the economic system seeks to continually produce more commodities for exchange, abstracting away from the socio-ecological effects of so doing (Foster et al. 2010; Paterson 2014; Miller 2008). The prioritisation of money, profit and exchange has links to the increasing financialisation of the economy. Carbon trading reflects this trend towards financialisation, because the carbon market enables actors involved in financial services to gain from brokering transactions in complex, abstract commodities (Layfield 2013; Fredriksen et al. 2014, 30–34).[1]

Use value has a wider reach than exchange value, because it refers to the wider purposes that things can hold. If for example, you use a carbon credit for decorative purposes, then it has an aesthetic use value. Furthermore, the carbon credit might impress your neighbours, in which case it also has some symbolic and status use values. If the purchase produces emissions reductions on top, then it has achieved an environmental use value. Likewise, should the offset produce some harmful consequences, then it can negatively affect some use value. The list of use values is therefore diverse and extensive, incorporating the many values that people potentially care about. As such, emphasising use value offers the possibility of a "broader ecological and social conception of wealth" compared to capitalism's narrower obsession with exchange value (Foster et al. 2010, 54).

Valuation processes, by contrast, refer to the actions involved in valuation, emphasising valuing as an action rather than value as a concept. Valuation processes can involve the measurement of use value to facilitate an object's market exchange. In the case of environmental markets, the goal is to produce commodities for exchange which incorporate some underlying use value. This incorporation is achieved through a valuation process, which is often fundamentally about measurement (Robertson 2012). For example, carbon offset commodities for exchange must be created through an often fraught process of valuation which seeks to measure supposed underlying use value, articulated in terms of quantities of 'emissions reduction'. In addition to measurement and quantification, valuation processes are also qualitative in nature, relying on cultural significations (Kjellberg and Mallard 2013). In the carbon offset market, significations of value are highly normative in character, as advocates construct (and critics deflate) an appearance of use value. This signification of value, as well as its measurement, will be explored further in the case of carbon offsetting. First, however, it is argued that the decline of offsets' exchange value is linked to the more fundamental failings of offsets to provide use value.

The declining value of offsets

At the time of writing, the exchange value of carbon credits generated in compliance markets (the CDM) has dwindled to almost nothing. Average prices for a Certified Emission Reduction (CER) credit from the CDM reached a high of €16.8 in 2008 (Kossoy and Ambrosi 2010), but in 2014 the average price for a CER was just €0.17 (Kossoy et al. 2015). There has been no significant price recovery since then. In effect, the market has collapsed due to the dramatic decline of the exchange value of its product. Some rationale for the extremely low exchange value of CDM credits is now offered (cf. Michaelowa 2014).

The primary source of demand for CERs and other credits from the United Nation's flexibility mechanisms has been the EU ETS. However, in recent years the EU ETS has been over-supplied with carbon credits. In 2014, emissions in the sectors covered by the ETS were already below the cap for 2020 (EEA 2015, 24). There were multiple causes of over-supply in relation to the ETS. First, the post-2008 European recession and economic slow-down reduced the amount of carbon intensive activity on the continent (Declercq et al. 2011). Indeed this reduction highlights the importance of planned economic de-growth strategies in richer countries, if climate targets are to be met in an internationally equitable manner (Anderson and Bows 2011). Second, the success of renewable energy technologies and other climate change mitigation efforts, such as energy efficiency targets, have reduced emissions in sectors covered by the ETS (EEA 2015). Third, the weak ambition of the ETS means that the emissions cap is set very generously, putting only a marginal constraint on the continuation of activities that cause climate change.[2] Similarly, the weak ambition of the Kyoto Protocol, including countries' widespread non-ratification of and non-compliance with the protocol, has constrained demand for carbon credits at the UN level. In turn, the weak restriction on carbon emissions in the ETS (and in the Kyoto Protocol) can be explained by the "political priority given to fossil-fuel-dependent capital", even "at the expense of the economic interests of carbon trading industries" such as the firms involved in offsetting (Bryant et al. 2015, 13). Fourth, the organisations regulated by the ETS have used a large amount of carbon offsets for compliance purposes, swelling the overall supply of carbon credits. Half of the overall emission reduction required under the ETS for the period 2008–2020 can be 'achieved' through the use of offset credits, which is around 1,600 million CERs (Alberola et al. 2015, 10). As the surplus of emissions allowances in the ETS is roughly 2,100 million carbon units, it is evident that the entry of offsets to the carbon market has greatly exacerbated the over-supply (Carbon Market Watch 2014). Thus, a combination of recession, alternative carbon reduction policies, weak targets and offsetting provisions has undermined the EU's flagship climate policy.

In response to the over-supply in the ETS, policy-makers at the European Commission (EC) focused upon a 'rescue package' for the carbon market, which has conditioned their approach to carbon offsetting. Policy-makers temporarily revived the carbon price by implementing 'backloading', a postponement of the supply of carbon credits for several years (Wettestad 2014).[3] Furthermore, policy-makers maintained quantitative restrictions on the amount of offsets that could be used for compliance purposes within the ETS, curtailing potential supply. The amount of CERs that can be used in the EU ETS is consequently restricted according to rules on 'supplementarity'. The third phase of the trading system, which runs from 2012 to 2020, has stricter rules on supplementarity such that 90 per cent of the total amount of carbon offset credits that can be surrendered by 2020 has already been used (Kossoy et al. 2015, 34; Kossoy and Guigon 2012, 21). Post-2020, the EU is aiming to remove offsets from the ETS entirely. Thus, over-supply of credits in the ETS has encouraged policy-makers to restrict the eligibility of offset credits in the world's biggest carbon market, thereby curtailing demand for these credits.

The initial rationale for including carbon offsets in the Kyoto Protocol and the EU ETS was to provide 'cost-effective' and 'flexible' emissions reductions so that the economic cost of climate change mitigation would be kept low (Girling 2010). For policy-makers, a key initial use value of offsets was thus to limit the economic cost of climate change mitigation (cf. Hovden and Lindseth 2004; Zannakis 2015, 228).[4] But as the carbon price declined and the credibility of Europe's premier climate policy was at stake, the low-cost use value dissipated and the intention of policy-makers was reversed. Their aim became to re-establish a higher carbon price in the EU ETS in order to improve the credibility of the policy. Restricting the eligibility of low-cost offsets in the EU ETS was a means of achieving those slightly higher prices and enhanced credibility.

Addressing the over-supply in the EU ETS was not the only reason that policy-makers moved to restrict offsetting. The original ruling on the 'supplementarity' quantitative restriction on offsets stems from scepticism among EU officials about the environmental use value of flexibility mechanisms

(Christiansen and Wettestad 2003, 11–12). Over time, concern about the quality of the credits, and a growing recognition that offsets regularly failed to produce the sustainable development and emission reduction use values intended, drove policy-makers to seek further qualitative limitations on offsetting. With mounting evidence for the drawbacks of offsetting and its lack of environmental use value, coupled with case studies highlighting the potentially negative developmental consequences associated with offset projects (e.g. Böhm and Dabhi 2009), policy-makers have had to re-evaluate the case for offsetting. A range of EU-commissioned reports studying 'the integrity of the CDM' briefed against the UN's flexibility mechanism, finding the CDM lacking both in its contribution to 'sustainable development' (du Monceau and Brohé 2011b) and in its environmental integrity (du Monceau and Brohé 2011a). The initial guiding rationales for implementing offsetting (i.e. producing the use values of cost-effective emission reductions and contributing to sustainable development), no longer retain the credibility they once had, as negative evidence about the problematic aspects of the CDM has grown stronger.

Qualitative restrictions on the eligibility of certain offset project types in the EU carbon market reflect policy-makers' concerns about the environmental use value of particular forms of offsetting. The EU has had long-standing restrictions on credits from the land use sector, namely afforestation/reforestation projects (Sterk and Kruger 2009). Land use projects only generate temporary CERs because it is not possible to guarantee the permanence of carbon sequestration achieved through biomass plantations (Thamo and Pannell 2015). This evident limitation in the environmental use value of land use projects encouraged the EU to exclude them. Other types of projects have also been exposed as having a similar lack in environmental value. The environmental integrity of the large quantities of offset credits derived from industrial gas projects (reduction of HFC-23 and abatement of N_2O from adipic acid plants) has been called into question (Schneider 2011; Schneider et al. 2010). In response, from the start of 2013 (EEA 2015, 36), these particular industrial gas credits have been excluded from the ETS. However, this comes rather late, considering that at least 550 million CERs from such projects have already entered the carbon market (Morris et al. 2013, 40). Overall there has been an increasing recognition that offsets across a broad range of project types, including the UN JI flexibility mechanism, frequently fail to provide the emissions reductions claimed (Kollmuss et al. 2015). The vast majority of offsets used in the EU carbon market are associated with problems of environmental integrity (Morris et al. 2013, 40–41). Arguably, the EU might have added more qualitative restrictions on offset project types, had the quantitative limit on offsets not already been 90 per cent fulfilled by the end of the second trading period in 2012.

The EU has put further qualitative restrictions on some offset project types because of concerns about their developmental use value. Large (above 20MW) hydropower offset projects entering the ETS are required to abide by the criteria for 'sustainability' set out by the World Commission on Dams as part of an effort to reduce the potential for harm associated with dams (Haya and Parekh 2011). Many other offset projects have produced demonstrable harms, but extra 'sustainable development' guidelines for these credits have not been introduced (Böhm and Dabhi 2009). Another regulation stipulates that, for the third ETS trading period (2013–2020), credits from projects registered with the CDM after a 2012 deadline will only be eligible if they are generated in least developed countries (LDCs). This regulation reflects that the use values of enabling 'sustainable development' and 'technology transfer' in developing countries (which motivated the introduction of offsetting at UN level in the late 1990s) have now met a very different political and economic context. Rapid manufacturing-led economic growth in China, India, Brazil and other countries which were the biggest hosts of CDM projects has led to a rethinking of the binary division of developed and developing countries in the Kyoto Protocol. Under Kyoto, only the former assumed mitigation responsibilities. The 2015 UN climate agreement in Paris expects that all signatories will commit to a reduction in emissions, giving rise to new complications for an offset mechanism because of the potential for 'double counting' the reductions associated with an offset (Schneider et al. 2014). If both the host developing country and the purchasing developed country claim to have reduced emissions from an offset project site, the principle of avoiding double counting will be violated. This helps explain the EU's decision to restrict eligibility of new credits to those from LDCs, which still have much less responsibility for reduction of carbon emissions compared to rapidly industrialising and historically developed countries. Furthermore, the decision to remove offsets entirely from the EU ETS from 2020 onwards will help to avoid potential double counting complications that would arise from the continuation of offsetting.

Thus, restrictions placed on offsetting in Europe's carbon market are related to policy-makers' enhanced appreciation of the lack of use value associated with offsets. Ending the eligibility of credits from the land use sector and industrial gas projects was prompted by concerns about the lack of environmental use value of those specific project types. The restriction of hydropower projects was aimed at enhancing their development value and avoiding harm. Restricting new projects from outside LDCs reflects the lack of political value in stimulating emissions reductions in rapidly industrialising countries, which are now expected to assume their own targets. Equally, the quantitative cap imposed on offsetting, which grew out of concerns about the probable lack of use value therein, has proved well founded. The cap has been maintained due to dissipation of the value of enabling low-cost emission reductions in the context of EU ETS over-supply. Perception of a lack of use value in offsetting has restricted eligibility, and put demand in decline, helping to explain the price crash.

Problems in the valuation process

The root cause of offsets' lack of use value lies in fundamental problems associated with the valuation processes involved in the creation of carbon offset commodities. The stated goal of offsetting is to measure emissions reductions and attribute them to an offset mechanism which issues certificates. Each certificate is intended to represent a reduction of one tonne of carbon dioxide equivalent. Thus, the certificate is an abstract commodity aimed at holding the use value of emission reduction. To create the certificate, the offset mechanism must establish a valuation process for it, through which the use value is imbued into the product. That valuation process needs to encompass both measurement and attribution, which for offsetting involves the particular challenges of constructing hypothetical 'baselines' and determining the status of a complex concept known as 'additionality'. These moves are highly problematic and will be considered in turn.

First, measuring the emissions supposedly reduced by an offset project requires postulation, and later verification, of a hypothetical 'baseline' level of emissions. The baseline is the emissions scenario for the counterfactual world in which the offset project has not yet occurred. The baseline is the reference scenario from which the actual emissions of the 'low carbon' reality are to be subtracted, to provide a quantified value of emissions reduction. As Larry Lohmann has recognised, offsetting thereby relies on a series of performative equations which includes, among others, the postulation that "actual CO_2 reduction = counterfactual CO_2 reduction" (Lohmann 2014, 174). Counterfactuals are necessary to the offset commodity because the latter's notional emission reduction value is equal to a counterfactual baseline, minus any actually measured emissions. The baseline scenario is a quantified conjecture about what the likely future emissions would have been in the absence of the offset project, and is therefore fundamentally uncertain. As Lohmann puts it, determining the baseline "demands the sort of knowledge that human beings have never attained, attempted, or believed possible" (2012, 99). Considering the above formula, the uncertainty of the postulated value for the baseline gives rise to an uncertainty surrounding the quantified value of the emission reduction.

Second, in terms of attribution, the valuation process involves navigation through the troublesome concept of 'additionality'. An offset mechanism must discriminate between those projects which simply reduce emissions, and those which reduce emissions because of offsetting. Where an initiative would have occurred even in the absence of the carbon offset mechanism, the said initiative is non-additional, and should not be considered an offset. The regulator for offsetting ought to aim for the registration of only those projects which can demonstrate that they owe their existence to the offset mechanism. Regulators faced with the task of determining the additionality/non-additionality of the offset have to grapple with difficult questions about causation. To some extent, this enters the same unknowable territory as in the determination of baselines, because postulations about what constitutes business-as-usual trends are indeterminate. Some form of decision must be made, however, which involves instituting practical additionality assessment procedures, all of which come with functional inadequacies (World Bank Group 2016; Schneider 2009).

The politics involved in the valuation processes which lead to both the definition of the baseline emissions and the attribution of additionality are such that baselines are likely to be inflated and additionality likely to be attributed, even when it should not be (cf. Chapter 7, this volume). With baselines, project developers can carefully select information which will convey to the auditor and the other regulatory bodies a picture of highly polluting future scenarios. The dirtier the baseline, the more emissions reductions can be credited, and the greater the revenue stream for the project developer. Economic interests in the acquisition of exchange value influence the 'calculation' of the baseline in an inflationary way that undermines the use value of the offset. With additionality, for project developers who prepare the documentation for review by auditors and the CDM Executive Board regulators, the economic incentive is equally clear: make the project appear additional. Project developers and their allies in consultancy work to present information in a carefully crafted way to give the appearance of additionality, according to the assessment tools in use. To do otherwise is to jeopardise the revenue stream.

Private sector auditors, known as Designated Operational Entities accredited under the CDM, are tasked with checking the information provided by the project developer on matters such as additionality and baselines, to confirm that the offset is in accordance with the methodologies, rules and procedures set by the CDM Executive Board. However, in the CDM the project developers are tasked with contracting the auditors, creating a conflict of interest whereby auditors have an incentive to be lenient with the presentation of 'facts' deployed by their clients. In an interview, one former auditor described this conflict of interest as an ongoing risk: "There is always a risk [of impropriety] because of the situation where we [auditors] are essentially a private company that is making our money from the clients that we are auditing" (cited in Watt 2016, 135). To retain business, auditors have an incentive to accept project developers' information, rather than spend time and effort seeking potentially contradictory third-party information. Kongsager and Corbera (2015) describe the case of a project developer taking action to replace an auditor who gave an unfavourable opinion on additionality with a different contractor, which illustrates the commercial politics involved in the assessment of this challenging concept.

Checks on project documentation submitted post-audit are in place at the next level up, as the CDM Executive Board examines project information on a case-by-case basis, sets rules for auditors and updates methodologies. This has the potential to encourage a reformist perspective, viewing the complexities involved in the valuation process associated with offsetting as 'technical challenges' to be addressed through improvements to methods of accounting (e.g. World Bank Group 2016; Schneider 2009). However, the problems of measurement are so far-reaching and substantial that technical reform efforts alone are unlikely to be sufficient. The problems of accounting for CO_2 in the processes of commodification necessary for carbon markets extend significantly beyond the problems of baselines and additionality mentioned here (Lohmann 2014). The valuation process necessarily involves the exclusion of relevant considerations in order to create a bounded commodity (Lohmann 2011, 2005). As Lohmann argues (2014, 178):

> With respect to the climate crisis in particular, internalizing externalities through commodity formation, however profitable the result, constantly gives rise to fresh externalities that are so overwhelming that, from an environmental point of view, they invalidate the project.

The flaws inherent in the valuation process for carbon offset commodities therefore act to produce a lack of environmental use value.

Changing evaluations of offsetting

One question which remains unanswered is to what extent the lack of validity in the carbon offset project leads to a decline in demand for offsets. Given the flaws of the system, what is called for is a series of 'deactivations' of the performative equations that produce carbon commodities, so as to render offsets ineligible for policy compliance and therefore undesirable in market exchange (Lohmann 2014). Policy-makers have now placed some restrictions on carbon offsets, showing that deactivation is under way. The core argument presented here is that evaluations of offsetting are gradually changing, thereby undermining the demand for carbon offset credits.

The section above showed the severe challenges of measuring and attributing carbon reductions under the CDM's politically and commercially sensitive valuation processes. As a result, users ought not to trust that offset credits hold the environmental value that is claimed. If users of offsets realised their lack of use value, offsetting might be considered as a market bubble at risk of crashing like the dot-com and other sectors in which underlying use value proved lower than expected (Layfield 2013). Indeed, the demand for credits among EU policy-makers has declined for this very reason, and the CDM has experienced a crash in exchange value. This case suggests that a normative element is at play in evaluations of offsetting, in so far as policy-makers are concerned to provide environmental value and social goods through their decisions.

However, the situation is more complicated than attributing the decline in exchange value to realisations about lack of use value alone. For one thing, the flaws of offsetting have been apparent to many observers since before the growth of the CDM (Lohmann 2005). Indeed, even at the time of negotiating the Kyoto Protocol, the EU's Environment Commissioner expressed severe doubts about the value of the flexibility mechanisms proposed by the United States (Cass 2005). So why the change now? Furthermore, it would be naïve to consider that policy-makers or other users of offsets act rationally or regulate based on information and the primary motives of environmental concern. Rather, policy-makers might be content to benefit from offsetting's use value of providing 'moral cover' for the continuation of fossil fuels combustion (Bachram 2004).

The most significant forms of evaluation for the compliance carbon market take place as policy debates, which are multifaceted and diverse, making generalisations difficult (cf. Zannakis 2015; Hovden and Lindseth 2004). To illustrate this, consider that while the EU has restricted offsets in the ETS, the use of significant amounts of otherwise restricted offsets is still allowed under the EU's Effort Sharing Decision (Holyoake and Poplawski-Stephens 2013). Unlike the EU's quantitative limits on the entry of offset credits into its carbon market, New Zealand's Emissions Trading Scheme has allowed unrestricted access to the UN flexibility mechanisms (New Zealand Environmental Protection Authority 2014; Harker et al. 2016). This diversity in rules calls for empirical investigations of the range of evaluations for carbon offsetting in policy-making forums, as well as in the voluntary market segment.

The broader issue that empirical studies should look to consider is the extent to which the carbon market can be sustained as a 'fictive economy' that lacks underlying use value, but which is constructed in narrative forms as valuable (Bracking 2015). Despite evidence to the contrary, advocates of offsetting, many of whom work in the carbon offset market and stand to gain from its sustenance, discursively frame offsetting as valuable in providing low-cost emission reductions (Descheneau and Paterson 2011). When pressed on offsetting's problems of value, these are framed as opportunities for improvement and reform, with insistence that offsetting is morally justifiable or even commendable (Watt 2016). The marketing of offsets, which includes publicity from the CDM, involves images of environmental protection and developmental care which are influential for perceptions of value.[5] These forces work to sustain the perceived value of offsetting, and aim to heighten demand and increase exchange value. On the other hand, critical commentaries, articles and media reports help to undermine the industry narrative, leading to recognition of the lack of use value in many credits.

What becomes clear is that public evaluations of the worth of carbon offsets, including in policy-making circles, are part of a broader valuation process that is distinctively normative in character. Efforts to stake out the moral high ground, conducted through discursive means, are part of the political–economic struggles involved in carbon trading. As Paterson describes: "legitimacy is becoming the condition of possibility of accumulation" in the carbon offset market (Paterson 2010, 361). A strong analysis of public moral evaluations of carbon offsetting is required to understand its valuation process.

Conclusion

This chapter has shown that differentiating between the three broad denotations of value – exchange value; use value; and valuation – facilitates an understanding of the major trends in the evolution of carbon offset markets. The analysis above indicates that the relation between use value and exchange value is mediated by valuation processes which operate in the realms of measurement, attribution and signification. To produce an environmental commodity like a carbon offset, one must measure some use value and attribute it to the commodity. But the scientific and political challenges of achieving this for carbon offsetting are profound, especially in a market context where many actors are pursuing commercial interests. The inability to imbue meaningful use value to the carbon offset commodity through the valuation process has led to a crash in the exchange value of these commodities.

However, as it is mediated through a valuation process of signification, this crash in exchange value was not guaranteed. The case of restricting offsets in the EU ETS shows that policy-makers came to have negative opinions about the worth of offsets in terms of their environmental, developmental, political and economic value. However, evaluations of offsetting in other policy contexts, or in the voluntary carbon market, can be quite different. Advocates of offsetting, many of whom have an economic stake in the market, still maintain that offsets are valuable. The discursive framings that advocates promulgate can be powerful enough to create a perception of value that is sufficient to sustain market demand. To counteract this, critical academic research, media reports and briefing papers can serve the important functions of challenging industry discourse and helping to realign public and policy-relevant evaluations of carbon offsetting. This should give a better appreciation of the lack of underlying use value of offsets, even if it leads to a crash in their associated exchange value.

Notes

1 Reflecting on trading volumes in the EU ETS, a World Bank report notes that 'a considerable portion of the trades is primarily motivated by hedging, portfolio adjustments, profit taking, and arbitrage' (Kossoy and Guigon 2012, 18).

2 By 2020, the EU aims to cut greenhouse gas emissions overall by just 20 per cent compared to 1990 levels (EEA 2015). However, research on carbon budgets shows that wealthy industrialised countries should be reducing emissions by around 10 per cent each year in order to make their fair contribution to reaching an internationally agreed 2°C target that avoids the most dangerous climate change (Anderson et al. 2014).

3 This weak backloading provision fell short of environmentally more stringent proposals to enhance the mitigation ambition of the ETS by permanently cancelling planned allocations, rather than postponing them. But even the meagre backloading proposal met opposition in the European Parliament, highlighting the political difficulty of implementing more fundamental environmentally ambitious reform of the ETS (Ervine 2014, 733).

4 The use of offset credits in the ETS is estimated to have reduced the monetary cost of compliance with the emission reduction targets for European organisations regulated by the carbon market by between €4 billion and €20 billion between 2008 and 2012 (World Bank 2014, 72).
5 The CDM has annual picture and video competitions which give overwhelmingly positive views of offsetting. The CDM has also established a means of enabling voluntary purchases of carbon offsets, conforming to some of the marketing techniques deployed on the voluntary market.

References

Alberola E., Afriat M., Rittenhouse K., Francis D., Sopher P., Deblock S. and Kouchakji K. (2015) *European Union: an emissions trading case study* Paris, Washington, DC and Brussels: CDC Climat, Environmental Defense Fund, and IETA.

Anderson K. and Bows A. (2011) "Beyond 'dangerous' climate change: emission scenarios for a new world" *Philosophical Transactions of the Royal Society A* 369: 20–44.

Anderson K., Le Quéré C. and Mclachlan C. (2014) "Radical emission reductions: the role of demand reductions in accelerating full decarbonization" *Carbon Management* 5: 321–323.

Bachram H. (2004) "Climate fraud and carbon colonialism: the new trade in greenhouse gases" *Capitalism Nature Socialism* 15: 5–20.

Böhm S. and Dabhi S. (eds) (2009) *Upsetting the offset: the political economy of carbon markets* MayFlyBooks, London.

Bracking S. (2015) "Performativity in the Green Economy: how far does climate finance create a fictive economy?" *Third World Quarterly* 36: 2337–2357.

Bryant G., Dabhi S. and Böhm S. (2015) "'Fixing' the climate crisis: capital, states, and carbon offsetting in India" *Environment and Planning A* 47: 2047–2063.

Carbon Market Watch (2014) *What's needed to fix the EU's carbon market: recommendations for the Market Stability Reserve and future ETS reform proposals* Carbon Market Watch, Brussels, http://carbonmarketwatch.org/wp-content/uploads/2014/07/ETS-POLICY-BRIEF-JULY-2014_final_1.pdf (Accessed 11 September 2017).

Cass L. (2005) "Norm entrapment and reference change: the evolution of the European Union position on International Emissions Trading" *Global Environmental Politics* 5: 38–60.

Cavanagh C. and Benjaminsen T.A. (2014) "Virtual nature, violent accumulation: the 'spectacular failure' of carbon offsetting at a Ugandan National Park" *Geoforum* 56: 55–65.

Christiansen A.C. and Wettestad J. (2003) "The EU as a frontrunner on greenhouse gas emissions trading: how did it happen and will the EU succeed?" *Climate Policy* 3: 3–18.

Declercq B., Delarue E. and D'haeseleer W. (2011) "Impact of the economic recession on the European power sector's CO_2 emissions" *Energy Policy* 39: 1677–1686.

Descheneau P. and Paterson M. (2011) "Between desire and routine: assembling environment and finance in carbon markets" *Antipode* 43: 662–681.

EEA (2015) *Trends and projections in Europe 2015: Tracking progress towards Europe's climate and energy targets*, European Environment Agency, Copenhagen, https://www.eea.europa.eu/publications/trends-and-projections-in-europe-2015v (Accessed 11 September 2017).

Ervine K. (2014) "Diminishing returns: carbon market crisis and the future of market-dependent climate change finance" *New Political Economy* 19: 723–747.

Foster J.B., Clark B. and York R. (2010) *The ecological rift: capitalism's war on the earth* Monthly Review Press, New York.

Fredriksen A., Bracking S., Greco E., Igoe J., Morgan R. and Sullivan S. (2014) *A conceptual map for the study of value* Leverhulme Centre for the Study of Value, LCSV Working Paper Series No. 2, Manchester, http://thestudyofvalue.org/wp-content/uploads/2013//11/WP2-A-conceptual-map.pdf (Accessed 11 September 2017).

Girling A. (2010) *NGOs and the Clean Development Mechanism: constraints and opportunities in the discourse of EU consultations* University of East Anglia, Governance of Clean Development Working Paper 5, Norwich, https://www.uea.ac.uk/documents/439774/5807661/GCD+Working+Paper+005+-+Girling+2010.pdf/08d2da8f-f070-436a-9600-7d15185674a5 (Accessed 11 September 2017).

Harker J., Taylor P. and Knight-Lenihan S. (2016) "Multi-level governance and climate change mitigation in New Zealand: lost opportunities" *Climate Policy* 17: 485–500.

Haya B. and Parekh P. (2011) *Hydropower in the CDM: examining additionality and criteria for sustainability* University of California, Energy and Resources Group Working Paper ERG-11-001, Berkeley, https://www.internationalrivers.org/sites/default/files/attached-files/haya_parekh-2011-hydropower_in_the_cdm.pdf (Accessed 11 September 2017).

Holyoake D. and Poplawski-Stephens P. (2013) *Discussion note: the offset scandals in the effort sharing decision* ClientEarth, Brussels, https://www.clientearth.org/reports/esd-and-international-offsets.pdf (Accessed 11 September 2017).

Hovden E. and Lindseth G. (2004) "Discourses in Norwegian climate policy: national action or thinking globally?" *Political Studies* 52: 63–81.

Kjellberg H. and Mallard A. (2013) "Valuation studies? Our collective two cents" *Valuation Studies* 1: 11–30.

Kollmuss A., Schneider L. and Zhezherin V. (2015) *Has Joint Implementation reduced GHG emissions? Lessons learned for the design of carbon market mechanisms* Stockholm Environment Institute, Stockholm.

Kongsager R. and Corbera E. (2015) "Linking mitigation and adaptation in carbon forestry projects: evidence from Belize" *World Development* 76: 132–146.

Kossoy A. and Ambrosi P. (2010) *State and trends of the carbon market 2010* World Bank, Washington, DC.

Kossoy A. and Guigon P. (2012) *State and trends of the carbon market 2012* World Bank, Washington, DC.

Kossoy A., Peszko G., Opperman K., Prytz, N., Klein N., Blok K., et al. (2015) *State and trends of carbon pricing 2015* World Bank, Washington, DC.

Layfield D. (2013) "Turning carbon into gold: the financialisation of international climate policy" *Environmental Politics* 22: 901–917.

Lohmann L. (2005) "Marketing and making carbon dumps: commodification, calculation and counterfactuals in climate change mitigation" *Science as Culture* 14: 203–235.

Lohmann L. (2011) "The endless algebra of climate markets" *Capitalism Nature Socialism* 22: 93–116.

Lohmann L. (2012) "Financialization, commodification and carbon: the contradictions of neoliberal climate policy" *Socialist Register* 48: 85–107.

Lohmann L. (2014) "Performative equations and neoliberal commodification: the case of climate" in Büscher B., Dressler W. and Fletcher R. (eds) *Nature Inc: environmental conservation in the neoliberal age* University of Arizona Press, Tucson, 158–182.

Michaelowa A. (2014) "Linking the CDM with domestic carbon markets" *Climate Policy* 14: 353–371.

Miller D. (2008) "The uses of value" *Geoforum* 39: 1122–1132.

du Monceau T. and Brohé A. (2011a) *Briefing paper "Baseline setting and additionality testing within the Clean Development Mechanism (CDM)"* AEA Consultancy, London, https://ec.europa.eu/clima/sites/clima/files/ets/markets/docs/additionality_baseline_en_0.pdf (Accessed 11 September 2017).

du Monceau T. and Brohé A. (2011b) *Briefing paper "Sustainable development and social equity: study on the integrity of the Clean Development Mechanism"* AEA Consultancy, London, https://ec.europa.eu/clima/sites/clima/files/ets/markets/docs/sustainable_development_en_0.pdf (Accessed 11 September 2017).

Morris D., Crow L., Elsworth R., MacDonald P. and Watson L. (2013) *Drifting toward disaster? The ETS adrift in Europe's climate efforts* Sandbag, London, https://sandbag.org.uk/wp-content/uploads/2016/11/Drifting_Towards_Disaster.pdf (Accessed 11 September 2017).

New Zealand Environmental Protection Authority (2014) *NZ ETS 2014 – Facts and Figures* Environmental Protection Authority, Wellington, www.epa.govt.nz/e-m-t/reports/ets_reports/annual/Documents/2014%20NZ%20ETS%20Facts%20and%20Figures%20final.pdf (Accessed 11 September 2017).

Paterson M. (2010) "Legitimation and accumulation in climate change governance" *New Political Economy* 15: 345–368.

Paterson M. (2014) "Commodification" in Death C. (ed.) *Critical environmental politics* Routledge, Abingdon, 53–62.

Robertson M. (2012) "Measurement and alienation: making a world of ecosystem services" *Transactions of the Institute of British Geographers* 37: 386–401.

Schneider L. (2009) "Assessing the additionality of CDM projects: practical experiences and lessons learned" *Climate Policy* 9: 242–254.

Schneider L. (2011) "Perverse incentives under the CDM: an evaluation of HFC-23 destruction projects" *Climate Policy* 11: 851–864.

Schneider L., Kollmuss A. and Lazarus M. (2014) *Addressing the risk of double counting emission reductions under the UNFCCC* Stockholm Environment Institute, Stockholm, https://www.sei-international.org/mediamanager/documents/Publications/Climate/SEI-WP-2014-02-Double-counting-risks-UNFCCC.pdf (Accessed 11 September 2017).

Schneider L., Lazarus M. and Kollmuss A. (2010) *Industrial N2O projects under the CDM: adipic acid – a case of carbon leakage?* Stockholm Environment Institute, Somerville.

Sterk W. and Kruger J. (2009) "Establishing a transatlantic carbon market" *Climate Policy* 9: 389–401.

Thamo T. and Pannell D.J. (2015) "Challenges in developing effective policy for soil carbon sequestration: perspectives on additionality, leakage, and permanence" *Climate Policy* 16: 973–992.

Tsing A.L. (2005) *Friction: an ethnography of global connection* Princeton University Press, Princeton, NJ.

Watt R. (2016) *The moral economy of carbon offsetting: ethics, power and the search for legitimacy in a new market* University of Manchester, Manchester.

Wettestad J. (2014) "Rescuing EU emissions trading: mission impossible?" *Global Environmental Politics* 14: 64–81.

World Bank (2014) *State and trends of carbon pricing 2014* World Bank, Washington, DC, http://documents.worldbank.org/curated/en/505431468148506727/pdf/882840AR0REPLA00EPI2102680Box385232.pdf (Accessed 11 September 2017).

World Bank Group (2016) *Carbon credits and additionality: past, present, and future* PMR Technical Note 13, World Bank Partnership for Market Readiness, Washington, DC, http://documents.worldbank.org/curated/en/407021467995626915/pdf/105804-NWP-PUBLIC-PUB-DATE-5-19-2016-ADD-SERIES.pdf (Accessed 11 September 2017).

Zannakis M. (2015) "The blending of discourses in Sweden's 'urge to go ahead' in climate politics" *International Environmental Agreements: Politics, Law and Economics* 15: 217–236.

9 Climate changing civil society

The role of value and knowledge in designing the Green Climate Fund

Jonas Amtoft Bruun

Introduction

In 2009, at the fifteenth international climate meeting 'Conference of the Parties (COP) to the UNFCCC in Copenhagen' (COP15), the UN Green Climate Fund (GCF) was conceived. The ambition was that the GCF would manage a large part of any future financial resources (including some funding diverted from other institutions) to assist developing countries in adapting to the impacts of climate change. The GCF was also mandated to catalyse finance for mitigation projects under various existing United Nations Framework Convention on Climate Change (UNFCCC) programmes and to attract investments from the private sector via a 'private sector facility' (GCF 2014b). A total of 130 Civil Society Organisations (CSOs) participated, as representatives of civil society and of those most affected by climate change, in the four-year GCF design process that took place from 2011 to 2015. This chapter explores the processes by which the activities of CSOs influenced the design and development of the GCF. It argues that two meta-discourses were prevalent within the GCF design process, namely *green governmentality* and *ecological modernisation.* It further proposes that these discourses worked in conjunction with each other to enforce an overall emphasis on financial and economic calculation, which acted to undermine alternative values presented by CSOs, in a third meta-discourse on *civic environmentalism* focused on human rights and pro-poor climate finance. The chapter contributes to the study of value by critically assessing how, in this particular case study of an emerging institution for managing climate finance and its internal calculative devices and technologies, valuation technologies in global climate governance were designed. It also illuminates the ways in which value calculations render some knowledge and services provided by CSOs relevant and acceptable, while others are discounted, thereby becoming subjugated knowledge.

Environmental governance and CSOs

A commonly repeated message from CSOs is that they are vital to international climate policy-making and implementation, to ensure the voices of those most vulnerable to climate change are heard by decision-makers (Bazán et al. 2008; Jasanoff 1997). CSOs are additionally considered to be essential in holding governments accountable for their pledges and for their responsibility to prevent dangerous anthropogenic interference with the climate system (Schalatek and Nakhooda 2012). Such perspectives are often heard and repeated in various forms in the halls and plenaries at international climate negotiations, particularly at the annual COPs of the UNFCCC.

The right of CSOs to participate in international environmental policy processes was established during the UN Conference on Environment and Development, also known as the Rio 'Earth Summit', first held in 1992 (Lousley 2015). Side events during the conference reflected an unprecedented openness to incorporating CSOs into global climate governance and the outcome document (Agenda 21) gave civil society a much more prominent role within environmental negotiations than in any other international policy processes (Carr and Norman 2008; Speth and Haas 2006; Lousley 2015). Including civil society in the framing of 'our common future' was part of the Brundtland Commission's (Brundtland et al. 1987) support-building strategy for the scaling-up of climate challenges from a matter of national concern to a global problem (Lousley 2015). Through a narrative of 'citizens of mother earth', the process successfully created 'global subjects' that would necessitate individual responsibility to avoid the changing climate and environmental decline (Lousley 2015; Foucault 1991). This was done through a series of international public consultations where civil society was invited to take part in the formation of solutions to the problem at hand (Brundtland et al. 1987). The stakeholders that contributed to the formation of ideas subsequently took an active role in pushing for implementation of the Commission's output and recommendations (Lousley 2015). The result for civil society was that CSOs were both empowered and invested in the climate policy regime that was forged at the Rio Conference in 1992, namely the UNFCCC. Public participation in decision-making and access to justice and information in environmental matters was further strengthened in the Aarhus Convention (UNECE 1998) and its Almaty Guidelines (UNECE 2005).

CSOs have worked diligently in the UNFCCC process, and interest among non-state actors has grown, such that there are often more observers at the UNFCCC COPs than there are state representatives (Winkler and Beaumont 2010; Nasiritousi et al. 2014; Hadden 2015). In one such COP, the 2009 international climate meeting (COP15) in Copenhagen, the UN GCF was conceived with the ambition to channel most future financial resources to assist developing countries in adapting to the impacts of climate change. The GCF was a particularly attractive political space to the CSOs for several reasons. First, as elaborated below, CSOs themselves were highly vocal in the call for a new climate finance institution during the Copenhagen Climate Summit (Abbott and Gartner 2011; Schalatek et al. 2011; Schalatek 2011). Second, in its constitutional document, the GCF was mandated by the COP to effect a paradigm shift in *inter alia* 'effective involvement of relevant institutions and stakeholders' (GCF 2011, 2). CSOs interpreted this to mean that they themselves were the relevant stakeholders. Third, after the failure of negotiations at the Copenhagen COP in 2009 (Hadden 2015), a number of CSOs saw the GCF as a 'more influentiable space'[1] and consequently directed high levels of effort and resources into the design phase of the fund (Wu 2013; Orenstein 2016a).

At time of writing[2] the GCF contains just under $10 billion and its governing board has accredited 33 institutions whose aim is to help disburse funding through climate finance projects. The projects already approved by the GCF are classified into three categories: adaptation (to respond to the consequences of climate change); mitigation (to reduce emissions of substances causing global warming); and mixed projects which pursue both objectives simultaneously. Examples of approved projects include respectively: building the resilience of wetlands; building climate-resilient infrastructure and enhancing early warning systems (i.e. adaptation); creating energy efficiency 'Green Bonds'; supporting 'off-grid' solar proliferation and finance opportunities for large-scale sustainable energy (i.e. mitigation). The CSOs have worked with the GCF in attracting finance as well as supporting the decision-making process by producing and inputting knowledge (Bracking and Dorsey 2015; Bracking 2014). Despite a high degree of agency awarded to CSOs in the design phase however, the GCF has come under heavy criticism from these same CSOs. They are now expressing 'a sense of sincere disappointment' (Orenstein 2016a, 1) with the GCF, indicated in statements about the institution such as that it 'does not look like the fund that civil society organizations had in mind when we fought so hard for its creation' (Wu 2015, 5).

The widespread disappointment among CSOs embedded in the process as 'insiders' triggers a set of interesting questions. First, it seems pertinent to ask: at what point during the process did the CSOs realise that the fund's design was going in the 'wrong' direction? Second, if the CSOs were aware early on that the fund was not going to be designed as hoped, why did they remain engaged? Third, did the CSOs believe that being embedded in the system would ultimately afford the opportunity to change the process from within? This leads to two final questions: (1) does the instrumental search for power to advance concepts of how climate finance should be governed (related to enhancing and protecting what CSOs ultimately value about ecology) necessitate assimilation to existing power structures? and (2) how do CSOs respond if such assimilation consequently displaces the very values and valuations systems they sought to advocate when joining? I will attempt to answer these questions in the remainder of this chapter (see also Bruun 2017).

Theorising civil society organisations

Research on civil society largely revolves around the ability of CSOs to improve the lives of 'the poor' and the capacity and legitimacy of CSOs to act as representatives for those without a voice in international policy processes (Pollard and Court 2008; Fisher 1997; Pallas and Urpelainen 2012). Another part of the literature challenges the ability of CSOs to offer convincing alternatives to mainstream practices and ideas about development, and argues that project-implementing CSOs are key vehicles for neoliberal policies (Hulme 2008; Uvin et al. 2000). This critique can roughly be divided between those that see a process of willing co-optation (Bebbington et al. 2008; Hulme 2008; Edwards 2008; Igoe and Kelsall 2005) and those that see CSOs as falling victim to tactics and the power play of hegemonic actors (Lohmann 2012; Bond and Dorsey 2011). There is also a third stream which upholds the possibility of CSOs resisting and remaining autonomous, while strategically engaging with hegemonic actors.

The CSOs engaged in climate change governance generally believe that their engagement holds the possibility of producing positive change for the poor and for mitigating and alleviating the harm caused by climate change – the third stream referenced above. However, they emerged disappointed from this design process. To understand the reasons for this, and to interpret whether CSOs may have been naively optimistic about what their contribution could offer, or whether they did indeed further their objectives, it is pertinent to review the form that their engagement took. CSOs essentially contributed to the design of key valuation systems and calculative devices, such as those used to evaluate projects and for gender, results-based management, expenditure tracking and safeguarding. Were the CSOs thus inevitably dragged into a neoliberal calculative universe beyond their control, or could they make a difference?

Green governmentality

Chapter 2 introduced the concept of governmentality and its importance to our shared conceptual map. In this case study it was used more specifically to analyse the power relations and institutional arrangements of the Intergovernmental Panel on Climate Change (IPCC) and the GCF design process. The CSOs join an institutional arrangement in which they largely remain a 'just invited' appendage. Climate change is formally governed by national decision-makers, usually high-level ministers or state leaders, that pursue multilateral negotiations within a big 'summit theatre' (Death 2011). Negotiators are informed by an eclectic epistemic community of scientists assembled in the IPCC, which ensures that climate-related decisions are based on environmental science (Jasanoff 2013). The IPCC draws its authority from global-scale observations of the climate system using satellite surveillance, computer-model simulations and 'paleoclimate reconstructions' (IPCC 2013, 2). Scientific analysis of its observations allows the IPCC to make predictions about the evolution of the planetary climate system within an 'acceptable error margin' (IPCC 2013). This science-centric regime has created a green governmentality in the locus of global climate politics, the

UNFCCC, which is underpinned on a global scale by carbon budgeting and measurement systems (Stripple and Bulkeley 2014; Purdon 2013). The 'green governmentality', constituted by these institutions, can also be viewed as having produced a discourse of climate governance.

Governing the climate in the form of a global carbon budget is part of this green governmentality, which uses various calculative technologies and devices (Blok 2011). Disparate sources of carbon are made commensurate through technical formulae (MacKenzie 2009) and results are monitored through a form of 'carbon panopticon' (Lövbrand and Stripple 2011). The knowledge that forms the basis for these carbon technologies often takes the form of a 'black box', in that it is neither contested nor widely understood (Latour 1999; Besel 2011). An example of a black box is found in the calculative formula used to turn various types of greenhouse gases into a tradable commodity in the form of a 'certified emission reduction' (CER). Although the process is not well understood among actors engaged with carbon markets, the empirical premise of CER is not challenged (MacKenzie 2009). This institutional arrangement conditions the dominant governance discourse of ecological modernisation (as we see below) which prominently features the challenge and consequences of an incomplete science.

Discourses of global climate governance

The CSOs work at the edges of this institutional arrangement in a universe of contested discourses, where discussions focus on arguments over which words or phrases should be added or deleted from text. International negotiations under the UNFCCC are characterised by endless hours of wrangling over wording and punctuation. Portions of text are taken in and out of brackets as countries seek consensus. Brackets are powerful indicators for what knowledge and agendas are subjugated. Likewise the text that 'loses its brackets' will be considered to represent acceptable knowledge and agendas. The most contentious policy items will often be the last to lose their brackets, usually in the early hours of the final night of a COP. In these wars of words, discourse takes a prominent role, and instils meaning to natural or social phenomena and problems and, thereby, the 'appropriate' policy tools to address them (Hajer 1995; Bäckstrand and Lövbrand 2006; Nielsen 2014; Litfin 1994; Li 2007). As Foucault explains, 'it is in discourse that power and knowledge are joined' (Foucault 1998, 100). Discourses are therefore characterised as groups of statements, which are not necessarily always homogeneous, as they may also include contradictions.

Such arguments over words and phrases are important because these items and concepts are themselves derived from (and indeed, generate) wider discourses and paradigms, through which power is ultimately made instrumental. An increasing number of scholars identify global climate governance as a political arena dominated by a number of simultaneously interlinked yet competing discourses that define what can be said and what should be left out of political discussions (Hajer 1995; Bäckstrand and Lövbrand 2006; Nielsen 2014; Litfin 1994; Lövbrand and Stripple 2011; Bäckstrand and Lövbrand 2016; Oels 2013, 2005; Methmann 2010). These authors have tracked how discourse is used to legitimise particular policies and programmes under the UNFCCC. Bäckstrand & Lövbrand (2006) show how case studies of UNFCCC programmes, such as REDD+ and CDM, can be treated as 'keyholes' through which broader climate governance can be observed.

Dominant discourses in climate change governance include what Luke (1999) has identified as a discourse of green governmentality, Hajer (1997) as a 'weak' version of ecological modernisation, and Bäckstrand and Lövbrand (2006) as 'reformist' civic environmentalism. A number of scholars have contrasted these discourses in various scholarly work (notably Oels 2005; Bäckstrand and Lövbrand 2006, 2016; Nielsen 2014; Methmann 2010).

Ecological modernisation

Building on this prior work, I advance the argument that the GCF represents a further strengthening of a neoliberal focus in the ecological modernisation discourse, entailing a move away from the science focus of green governmentality and the stakeholder participation focus of civic environmentalism. Ecological modernisation is a dominant narrative asserting that climate research will always be flawed by scientific uncertainty, which means that the more the IPCC expands its research, the higher the level of uncertainty will become (Methmann 2014). An example of this scientific uncertainty is found in the attempts by the IPCC to establish the causes and impacts of climate change. Although the IPCC has now established that climate change is with 95 per cent 'certainty' manmade, uncertainty about the timeline and impacts on the local, regional and global scale remains (Purdon 2013). The absence of completely incontestable scientific documentation at the UNFCCC has fostered a precautionary principle adopted in several major UN conventions (Speth and Haas 2006). This norm states that 'lack of full scientific certainty shall not be used as a reason for postponing cost-effective measures to prevent environmental harm' (Speth and Haas 2006, 91). The idea that climate measures should neither be postponed, nor 'unduly costly' (Speth and Haas 2006, 91), has given room to a dominant advanced liberal governmentality of ecological modernisation that argues that climate policies must be justified by economic rationalities, characterised by narratives on 'sustainable development' and 'green growth' (Andersen and Massa 2000; Hajer 1995; Brand 2010).

Anja Oels (2005) explained that ecological modernisation is the green version of advanced liberalism, which understands markets as the optimal way of structuring societies, and holds that climate protection must therefore be performed through flexible and cost-optimising mechanisms. This is characterised by the flexibility mechanisms of Kyoto, such as carbon markets and the CDM.

> Ecological modernisation believes that a free market setting and limited government incentives will spur technological innovation that solves the ecological crisis in a cost-efficient manner.
>
> (Oels 2005, 196)

Although green governmentality and ecological modernisation are seemingly dichotomous, it has been shown that these discourses can be mutually reinforcing. In fact, one discourse can be captured by another and made to support a particular field of knowledge (Bäckstrand and Lövbrand 2006; Oels 2005). The two discourses overlap in several areas including in their focus on global rather than local or national solutions, and the reliance on somewhat complicated technologies from the natural sciences as a way to legitimise policies to the broader public (Bäckstrand and Lövbrand 2006, 2016). Tobias Nielsen (2014) perceives a 'marriage of convenience' between the two in which ecological modernisation provides technical and market-based solutions, which depend on CERs and other technologies, however both ultimately rely on 'the market' to resolve things. Hence, they are not mutually exclusive, but can thrive on each other. Carbon markets need scientific monitoring, verification and certification to be a viable climate solution. This is provided by the green governmentality discourse, which has 'provided the scientific and administrative rationale for measuring, monitoring and certifying carbon removals. [...] In contrast, ecological modernisation operates as a legitimising discourse – a blueprint for action' (Bäckstrand and Lövbrand 2006, 68–69). The 'climate problem' is thus framed in a power/knowledge nexus, which renders particular problem-solving techniques and the associated actors axiomatic, while others are rendered esoteric (Foucault 1991; Gane 2008; Dean 2010). Researchers have exposed routine practices of problematisation under the auspices of the UNFCCC based on predetermined intelligibility, expertise and intervention-formulas in the CDM (Oels 2005; Methmann 2011), forest protection (Bäckstrand and Lövbrand 2006), carbon markets (Lovell and Mackenzie 2011) and green bonds (Bracking 2015a). A common proposition in this research is that framing the nature of problems in terms that predetermined solutions and actors can address is a highly powerful exercise. This is because it ensures relations of self-legitimising power, in that the agents that emerge as necessary are often perceived as impervious experts, and the solutions are labelled 'apolitical'. Neither can be objected to despite operating in sensitive political terrains (cf. Li 2007).

In the GCF, this problematisation has taken the form of 'financialisation', which argues that the climate problem is inherently financial and can only be solved by adding more capital in the form of climate finance (Bracking 2015b). This is reinforced by a narrative on 'efficiency' (framed as 'value for money'), which is the notion that mitigation should take place, temporally and spatially, where it will have the highest impact for the lowest monetary cost (also see Fredriksen, Chapter 3 in this volume). Despite the historic responsibility for emissions causing climate change lying largely in the global North, this has been diluted by the financialisation of policy which presents a more global framing of problematisation and responsibility. For example, the CDM has created an economy for 'responsibility outsourcing' to developing countries, resembling a form of carbon colonialism (Bumpus and Liverman 2008). In the GCF this logic takes the form of 'cost-effectiveness and efficiency regarding financial aspects' (GCF 2015a, 29) (framed as non-cost optimising), which can automatically exclude spending on development.

Civic environmentalism

Alongside the green governmentality and ecological modernisation discourses exists a third discourse, civic environmentalism, which has both a radical and a reformist component. Reformist civic environmentalism argues for a 'modern' form of capitalism where external impacts on the climate are internalised in the capital system. As such, this version is closely aligned with ecological modernisation in its focus on reforming, rather than abandoning, capitalism (Bäckstrand and Lövbrand 2006). Conversely radical civic environmentalism (dominant among social movements that choose not to participate in, but protest against, the COPs) fundamentally questions the global capitalist system and argues for a restructuring of the growth-oriented paradigm dominating the global economy. The radical civic environmentalism discourse has formed the conceptual basis of the 'climate justice movement', which has won traction with CSOs on the inside of COPs (Bäckstrand and Lövbrand 2016).

Both components of civic environmentalism revolve around enhancing stakeholder engagement in global climate governance. Many believe that distinguishing between the two is complex as, arguably, the counterposed camps of 'reformist' and 'radical' are not entirely distinct (Bäckstrand and Lövbrand 2006, 2016; Nielsen 2014; Methmann 2010). However, differentiating between the two strains is useful for the purpose of this chapter. This is because many CSOs would arguably consider themselves to be both 'radical' and be acting 'radically' in the UNFCCC process through collective walkouts (e.g. Wu 2013; Hadden 2015), while remaining reformist in the GCF process by *inter alia* refraining from attempts to obstruct the process (Bruun 2017). The discrepancy between how the CSOs actually spoke and acted in the GCF will be further explored in the next section.

The GCF

These discourses on environmental governance inform how climate finance is viewed and the institutions which spend it designed. In the following section this is exemplified by the design process of the GCF, which emerged at a nadir in the history of climate finance and its institutions. This was because the central institution which preceded it – the CDM – was a market-based mechanism of climate finance inspired by ecological modernism. By 2009, when the GCF was first advocated at the Copenhagen COP, the CDM was in economic and moral decline (cf. Chapter 8, this volume). Several high-profile cases of human rights abuses and violations of social, gender and environmental performance standards had been brought, such as one in which the right of affected communities to be consulted before a CDM project was approved (Bond 2012; Carbon Market Watch 2014b; GAIA 2016; Olsen 2007; Munnings et al. 2016). Particularly problematic cases included the building of dams in the Barro Blanco Hydroelectric Plant in the Western Panama river of Tabasará and the Santa Rita Hydroelectric Plant in the Dolores River in the Alta Verapaz region in Guatemala (Both Ends 2016; Carbon Market Watch 2014b). Both were approved as CDM projects despite strong opposition from affected communities and both were constructed with co-finance from international finance institutions, including the World Bank through the International Finance Corporation (IFC 2016) and Dutch Finance Company for Development (FMO). The construction of these dams has been justified by their alleged climate benefits (IFC 2016), but has not undergone appropriate consultation and is allegedly limiting local communities' access to water, food and freedom of movement. Furthermore, protesting communities were met with violent force by the Guatemalan government 'which resulted in severe human harms, including the alleged deaths of seven people, amongst them two indigenous children of 11 and 13 years' (Carbon Market Watch 2014b, 1).

The environmental and social demise of the CDM is entwined with its economic decline. The price of CDM projects has plummeted, leading the CDM executive panel to conclude that the CDM had essentially collapsed (CDM 2012). CERs traded at less than $1/ton in February 2016 (Vertis 2016). Relatedly, much research has indicated that the CDM has done little in terms of actual climate protection (Bond 2012; Munnings et al. 2016; Olsen 2007; Watt 2016). In fact, money has been shown to flow towards fossil fuel projects claimed to be 'less' polluting (Bracking 2015b) and to non-fossil investments, such as hydropower, both of which have been met with suspicion regarding their climate benefits (Flyvbjerg 2005; Ansar et al. 2014; Oxfam et al. 2015; Huber and Joshi 2015; Franks et al. 2014).

Due to this falling price curve and together with increasing protests from affected communities and social movements, CDM projects are no longer instigated. This has entailed an extensive lack of important funding flows for mitigation and adaptation in developing countries (Bond 2012). This has *inter alia* left the Adaption Fund (resourced by a 2 per cent levy on CDM projects) with a large backlog, which it is unable to fund (Bracking 2014). Although developed countries agreed to put $100 million in the Adaptation Fund during COP18 in Doha, this money covered the backlog up until 2012 only, and the Adaptation Fund has since been only sporadically replenished (The Adaptation Fund 2016). The demise of the Adaptation Fund is seen by developing countries as highly problematic, because it was the only large source of public adaptation funding that could be accessed directly through national implementing entities (provided agreed fiduciary standards are met), as opposed to going through multilateral development banks or UN agencies (Caravani et al. 2012). Due to problems with the CDM, the dismal state of the Adaptation Fund and the wish for greater ownership of climate finance, a separate institution for the delivery of climate finance was requested in the last days of COP15 in Copenhagen (Corbera and Brown 2010; Brown et al. 2013). The strongest request came from Mexico, which led a coalition of developing countries and CSOs from the global North and South.

The GCF was presented as a paradigm shift in the delivery of climate finance, which entails ownership by developing countries, inclusion of stakeholders and direct access to finance (GCF 2015b). These aspects are important in the discourse of civic environmentalism. Several years after Copenhagen, it still holds many promising features, such as an equal governance structure evenly split between developing and developed countries, a strong focus on delivering climate finance in a 'gender sensitive' manner and the inclusion of recipient countries in both decision-making and implementation (Rowling 2014; Bruun 2016). However, critics argue that the GCF process has been anything but inclusive to stakeholders (Wu 2013; ActionAid USA et al. 2015) and so far very few developing countries have obtained direct access to funding. This is instead being disbursed through a 'who is who' roster of multilateral development and international commercial banks (Schalatek et al. 2015; Orenstein 2016a; Wu 2015; Redman and Orenstein 2015; Bracking and Dorsey 2015). How the political climate around the GCF has oscillated between towering hopes and fundamental despair will be explained in the following section.

Designing the GCF

The GCF was presented as something that would be different from existing market-based and non-market based climate finance institutions. However, actors had profoundly divergent views about how the difference would materialise. The disagreements were largely split along the lines of the two competing discourses (ecological modernisation and civic environmentalism), with contentious debates over who the GCF would serve, how resources would be mobilised and how funding would be disbursed from the beginning of the design phase (Bracking 2014). For ecological modernists it was also a rearguard action to prevent too much democracy and ideas of climate justice entering the debate. The data in this section are based on: (1) a content analysis of the technical documents submitted in the 2011 design phase; and (2) over 50 semi-structured interviews with relevant actors, conducted between 2012 and 2016 at GCF Board Meetings 3, 5, 8 and 9 and at the UNFCCC's annual meeting for its members (Conference of the Parties), COP 18, 19, 20, 21 and 22. The research approach was based on Collaboratory Event Ethnography (Campbell and Brosius 2010; Bracking 2014).

The GCF design phase started in 2011 when a Transitional Committee (TC) was tasked with creating what would be a governing instrument for the GCF (Schalatek & Nakhooda 2013). The TC awarded a high level of agency to CSOs, which were invited to contribute with expertise and submit documents for preparation and decision to the Technical Support Unit (TSU) (Schalatek et al. 2011; Schalatek 2011). The TSU acted as facilitator to the 40-person board by preparing decision papers, answering technical questions and synthesising knowledge and information from stakeholders for further elaboration. CSOs and other non-state stakeholders were permitted to sit in the same room as the decision-makers and to make comments during discussions. In line with the discourse of civic environmentalism, stakeholders were invited to be part of the policy discussions during the TC which looked promising for the CSOs participating in the GCF process. The high level of accessibility meant that CSOs were successful in introducing particular terminology into documents of the TC members (notably a redress mechanism for communities negatively affected by a GCF project and mainstreaming gender considerations into the context of the GCF's work). Also linguistic signifiers composed by civil society such as 'country-owned', 'participatory and inclusive processes', 'transformational change', 'long-term' and 'paradigm shift'[3] found their way into the Governing Instrument (GCF 2011). This large degree of influence meant that CSOs became highly professionally invested in the fund.

During the early design phase, it was possible to see two opposing models at play, roughly corresponding to the discourses of ecological modernisation and civic environmentalism. In the former (mostly advocated by developed country board members and the private sector), the GCF was depicted as a bank or development finance institution separate from the UNFCCC. Here the GCF was seen to offer complex financial instruments and use public funding to leverage private sector investments through the usual range of institutional partners in the development finance architecture, and a voting model based on contributions, similar to the World Bank or IMF (Bracking 2014). Conversely, in the civic environmentalist model (mostly advocated by CSOs and developing country board members), the GCF was depicted more like a public aid agency closely tied to the UNFCCC. In this reading, the GCF was seen to offer domestic mitigation from developed country members who would also provide grants and highly concessional public funds to adaptation and mitigation in developing countries, using national entities from developing countries (thus avoiding the bad track records of international intermediaries, such as Multilateral Development Banks (MDBs), UN agencies and international NGOs, in respect of environmental and social violations in the Climate Investment Funds (CIFs), Global Environment Facility (GEF) and CDM (CAN 2011, 2013; Harmeling et al. 2013; Wu 2015; Nacpil et al. 2015; ActionAid USA et al. 2015)). Voting would be distributed to all members equally and participation from stakeholders, civil society and affected communities would be encouraged.

However, when it came to operationalising the Governing Instrument through a Business Model Framework (BMF) for the fund, the ability to input knowledge to decision papers became increasingly restricted for CSOs. Several logistical features changed when the TC was replaced with a permanent GCF Board and the TSU was replaced with an interim secretariat in 2012. The CSOs were no longer allowed to sit in the same room as decision-makers, and a limit of only one active observer each from a northern and southern CSO were permitted to attend. These persons would only be invited to speak after the board had finished discussing an agenda item. The interim secretariat was eventually replaced with an incrementally larger permanent GCF secretariat. Some of the staff from the TSU and the interim secretariat carried through to the permanent secretariat, while others decided not to move to Songdo in Korea where the fund is located. Due to low levels of core staffing, the interim and permanent secretariat needed outside consultants to write decision papers that would form the basis for board debates on the GCF structure, while some were temporarily seconded to the GCF. These came from international management consulting firms, such as McKinsey & Company, PricewaterhouseCoopers, a number of retirees from the World Bank and other MDBs and also research-oriented CSOs such as Overseas Development Institute, Climate Policy Initiative and World Resources Institute (GCF 2015c). Thus, in contrast to the TC process, where a wide variety of consultants from the private sector, civil society and existing institutions were used to feed in diverse knowledge or opinions (Schalatek and Nakhooda 2014), by the time of the GCF implementation phase 'you had a preponderance of white older males with a background coming from the World Bank and the IFC'[4] that seemed to have been 'handpicked by the secretariat'.[5] These consultants drew on their background experience at multilateral banks, the World Bank and IFC, and from literature reviews generated within these organisations, to produce a certain type of finance-oriented knowledge, path dependent on existing practice and embedded in financial logics (Müller 2013).

Due to the move toward financialised calculation, the knowledge that the CSOs had to offer in the GCF process swiftly went from recognised to subjugated knowledge. The term *recognised knowledge* in this context refers to knowledge which is rendered relevant in papers by the secretariat, while the term *subjugated knowledge* refers to knowledge which is rendered irrelevant or unconstructive. This falls under the concept of creating *intelligent fields* (Dean 2010; Li 2007). In other words, while the TC TSU was looking for experience in designing a UN fund with a focus on stakeholder participation (something of which the CSOs had proven knowledge), the GCF secretariat had hired 'retirees from the World Bank',[6] who only paid attention to a sort of expertise that the CSOs could not offer,[7] such that the multi-stakeholder consultations pioneered by the TC were largely removed from the GCF implementation process.[8] It was still possible for CSOs to submit inputs, while decision papers produced by the secretariat would often assert that civil society had been consulted.[9] However, sources felt that it was impossible to identify if or where their views had been taken on board in the final papers (for this point see also Orenstein 2016a).

The relative disempowerment of CSOs allowed consultants to prepare decision papers on the BMF, grounded in disputed science, which framed the climate problem in discursive terms from within the paradigm of ecological modernisation (Bracking 2014). A principal decision paper titled 'Business Model Framework: Objectives, Results and Performance Indicators' (GCF 2013a) prepared by consultants from McKinsey & Company is indicative of how technical decision papers restricted debate by framing the options available to the board. Released close to the fourth board meeting in Songdo, June 2013, the paper chose 15 'priority result areas' for GCF funding. These covered mitigation, adaptation, and the hybrid definition of the two introduced in the Governing Instrument as 'cross-cutting' (GCF 2011). All 15 priority areas were claimed in the paper to be 'fully comprehensive' (GCF 2011, 3), but presented a constricted focus on mitigation potentials in small-scale agriculture, combined with the designation of most 'adaptation' areas in terms of service goods, such as 'knowledge hubs' (GCF 2011, 5) and 'research and development' (GCF 2011, 5). Omitted were infrastructural assistance to affected communities, such as low-cost housing and water management support, which suggested a narrowing of the field of possibilities from which the board could choose.

The paper established an implicit hierarchy between climate expenditure and sustainable development. This is epitomised by two counterposed strategic models that the board members were asked to choose between. Option one, 'strategic focus on climate mitigation and adaption' (GCF 2013a, 4), would, according to the paper, 'raise the prospects for effectiveness in achieving outcomes' (GCF 2013a, 4), despite giving 'less priority to country ownership' GCF 2013a, 4). Meanwhile, option two, 'strategic focus on maximizing co-benefits through strong synergy with sustainable development' (GCF 2013a, 4), would, 'likely ensure a country-driven approach', but would be 'less effective in meeting other objectives' than option one and thus limit a 'focus on direct mitigation and adaptation benefits'. Faced with the choice between climate expenditure and supporting sustainable development, the board adopted the former.

The debate around this paper was highly contentious, particularly because the consultants had used the 15 priority result areas in most other BMF papers before they had been discussed by the board or subjected to scientific peer-review by environmental scientists. This was problematic, because the instrumental repetition of non-scientific and non-consensus areas appeared as an attempt to imbue them with scientific and political authority and solidification. As a result, developing country board members were vehemently opposed to having priorities forced upon them as follows:

> The authors of this paper do not have the mandate to set the mitigation priorities for developing countries and nor can the board adopt these

proposed priorities without reference to what has been agreed to under the convention.[10]

Moreover, the CSOs opposed the paper. In an intervention to the board they argued that the priority results areas seemed to have 'emerged out of a black box'[11] in that they were grounded in 'questionable science',[12] had not been prepared through a multi-stakeholder engagement process, and blatantly attempted to gloss over existing controversies and arrest debates, which had not been solved at the GCF or even UNFCCC.[13]

However, despite profound disagreements, at its fifth meeting in Paris, in October 2013, the board approved a BMF for the GCF closely aligned with the one presented in the decision paper (GCF 2013b). The version of the BMF presented in this controversial paper emerged due to a combination of practices throughout the design process. Indeed, while the use of long and complex decision papers was exemplified above, the decision was conditioned by the three other practices delineated at the start of this chapter: 'consensus-by-exhaustion', labelling decisions initial, and linking financial promises to policy progress.

The decision on the BMF was taken at the end of three long and exhaustive days and rushed through in the last hour under the premise that the GCF was likely to remain empty if a decision was not taken at that meeting.[14] In fact, board members from developed countries[15] linked contribution to the fund with progress on the BMF, while asserting that policies were merely 'initial and indicative'[16] and 'could be refined'[17] at a later point. Several developed countries claimed that 'taxpayers'[18] needed to see the 'animal'[19] before they could start talking about putting money into the fund. 'There can be no pledging session before the business model framework is in place.'[20]

Although decisions on the priority areas were labelled provisional, they ultimately provided the sole basis for future documents and decisions on where the GCF would focus its investments. An indication of this is the subsequent 'Results Management Framework' (GCF 2014b) and 'Investment Framework' (GCF 2014a), which have both been based on the same 15 'initial' priority result areas. Thus, we can trace how technical decision papers, exhaustive negotiations, labelling long-term decisions as initial, and the threat of resources (or lack of) were used to govern the debate and decisions on what the GCF will fund and how funding will be disbursed.

In sum, the GCF design process was characterised by a move from civic environmentalism towards a dominant advanced liberal governmentality of ecological modernisation, characterised by the problematisation of the climate as one that is purely financial, with technologies of valuation of projects, interventions and results de-coupled from any validated environmental science. Interestingly, it seems that the discourse of ecological modernity became so hegemonic that it did not 'need' green governmentality to provide scientific legitimacy for spending priorities. Instead it seems that repeating messages in multiple papers was enough to assert a level of knowledge and scientific authority in the absence of actual science data, a point made elsewhere by Sullivan (2016).

After the 2013 BMF debate, and the decline in CSO influence that it signalled, participating CSOs and observers changed their focus. Rather than solely attempting to affect what the fund would do, they focused on injecting language on what the fund should *not* do and what measures of redress should be taken in the event that the GCF funded a project causing environmental or social harm (Redman and Orenstein 2015; Carbon Market Watch 2014a). More specifically the CSOs lobbied the board to adopt an investment 'exclusion-list' of what the GCF should not fund, blocking accreditation of entities that may be particularly 'un-attractive' to civil society, such as Deutsche Bank, HSBC and Credit Agricole (Carbon Market Watch 2014a; Orenstein 2016b).

While the board has continuously refused to adopt an exclusion list or block any entities from accreditation, this issue again demonstrated that the CSOs concerned remain curiously committed to their engagement with the board and are not willing to withdraw (either partly or wholly) from the discussion processes. Rather, they move the parameters for success or redefine their own. Thus, CSOs celebrated a decision on re-accreditation (rather than exclusion) which dictates that an initially unsavoury (but still accepted) entity must demonstrate a 'portfolio-shift' to obtain re-accreditation after five years. For example, an entity funding fossil fuel-based sectors would have to fix an initial baseline, and show under the Monitoring and Accountability Framework of the GCF that they have shifted their portfolio away from climate-damaging activities as a requirement for re-accreditation after five years. How far this incentivises fossil fuel-based investors to move to cleaner technologies however remains to be seen.

Conclusion

This chapter has argued that the design phase of the GCF lends itself to an interesting case study of how, in order to enhance their influence in international policy processes by the achievement of small and incremental changes, CSOs may abandon their activist ontology of promoting ideal normative positions while opposing imperfect policy frameworks. In this strategy, or form of engagement, the CSOs encountered the micro-detail of calculative technologies and calculative entities for evaluating the value and worth of competing environmental interventions, policies and investments. Thus, this chapter has contributed to the study of *value* by critically assessing how valuation technologies in global climate governance were designed at a salient moment (i.e. at the birth of the GCF), and examining the ways in which value calculations were performed that rendered some knowledge and services provided by CSOs relevant, while others became subjugated.

In their engagement with the institutional design of the GCF, the CSOs were working in a terrain of contested and competing discourses and paradigms which made sense (or not) of their advocacy. I argued that two meta-discourses prevalent within the UNFCCC process, green governmentality and ecological modernisation, worked in conjunction to enforce an economic calculus in the GCF design, which undermined values presented by the CSOs on human rights and pro-poor climate finance captured in a discourse on civic environmentalism. I demonstrated how the CSO lobby was undermined throughout the design phase in matters pertaining to spending objectives, as its knowledge and services were subjugated to this higher order financialised 'science', while at the same time winning incremental victories on gender policy. The lack of indisputable science regarding climate change made it possible for the pro-business lobby to remove priority areas from scientific benchmarking, while asserting that climate policies must be made commensurable with and even contribute to economic growth.

The GCF design phase essentially saw CSOs abandon their holistic and global scale analysis, to provide highly technical knowledge inside the realm of financial methodologies. Arguably, this metamorphosis of climate advocacy from purist normativity to strategic engagement is not sufficient to predict a loss of influence. However, it happened in the context of, and in response to, the 'financialisation of global climate governance', when mainstream climate governance became increasingly dominated by neoliberal policy agendas, which entrench financial methodologies and valuation processes and problematise climate change as a matter of attracting more 'green' investments (Sullivan 2013; Bracking 2014; Bracking 2015b). This makes the opportunity cost of deepened CSOs' involvement in the GCF particularly high, in terms of whether CSOs can win even small changes in such an inhospitable terrain. Additionally, these costs rise further because the cost of engagement must also include the problematic effect of how far CSO involvement legitimises governance processes that are deficient in protecting the climate and/or delivering global development. In this respect, CSOs risk devaluing their brands and reputation in global climate finance.

Notes

1 Interview with large US-based CSO (source 5) at the 21st Conference of the Parties, Paris, December 2015.
2 July 2017.
3 All signifiers are derived from CAN-International comments on 'Draft report of the Transitional Committee' of 7 October Submission to: Transitional Committee of the Green Climate Fund (14 October 2011).
4 Interview with large Washington-based CSO (source 1) at the eighth Board meeting, Barbados, October 2014.
5 Interview with large Washington-based CSO at the (source 4) ninth Board meeting, Songdo, March 2015.
6 Interview with large London-based CSO (source 3) at the eighth Board meeting, Barbados, October 2014.

7 Interview with large Washington-based CSO (source 2) at the eighth Board meeting, Barbados, October 2014.
8 This is according to CSO representatives. This section is based on data from field-notes.
9 Decision papers will have a version of the following section under 'approach': [This paper is] 'Based on an extensive literature review and completed with inputs and guidance from representatives of United Nations Framework Convention on Climate Change (UNFCCC) secretariat supporting the thematic bodies, Board members, as well as experts from civil society, think tanks and the private sector' (GCF 2015a).
10 Mr Omar El-Arini (Egypt) quoted in Raman (2013c).
11 CSO active observer intervention at the fourth Board meeting.
12 CSO active observer intervention at the fourth Board meeting.
13 CSOs' intervention bullet points circulated on the CSO listserv in June 2013.
14 Event ethnographic fieldwork at the fifth Board meeting in Paris, October 2013.
15 Notably UK, Australia, Norway and the US.
16 Mr Nick Dyer (UK) quoted in Raman (2013b).
17 Mr Alexander Severens (US) quoted in Raman (2013b).
18 Several developed countries referred to need for GCF passing the 'tax-payer test'.
19 Mr Jan Cedergren (Sweden) quoted in Raman (2013a).
20 Mr Rod Hilten (Australia) quoted in Raman (2013a).

References

Abbott K.W. and Gartner D. (2011) *The Green Climate Fund and the future of environmental governance* (Working Paper No. 16) EARTH System Governance, Lund.

ActionAid USA et al. (2015) CSO's open letter to GCF Executive Director on Stakeholder Participation. Songdo.

Andersen M.S. and Massa I. (2000) "Ecological modernization—origins, dilemmas and future directions" *Journal of Environmental Policy & Planning*, 2: 337–345.

Ansar A. et al. (2014) "Should we build more large dams? The actual costs of hydropower megaproject development" *Energy Policy*, 69: 43–56.

Bäckstrand K. and Lövbrand E. (2006) "Planting trees to mitigate climate change: contested discourses of ecological modernization, green governmentality and civic environmentalism" *Global Environmental Politics*, 6: 50–75.

Bäckstrand K. and Lövbrand E. (2016) "The road to Paris: contending climate governance discourses in the post-Copenhagen era" *Journal of Environmental Policy & Planning*: 1–19.

Bazán C. et al. (2008) "Producing knowledge, generating alternatives? Challenges to research-oriented NGOs in Central America and Mexico" in Bebbington A.J., Hickey S. and Mitlin D.C. (eds) *Can NGOs make a difference? The challenge of development alternatives* Zed Books, London, 175–196.

Bebbington A.J., Hickey S. and Mitlin D.C. (2008) "Introduction" in Bebbington A.J., Hickey S. and Mitlin D.C. (eds) *Can NGOs make a difference? The challenge of development alternatives* Zed Books, London, 3–38.

Besel R.D. (2011) "Opening the 'black box' of climate change science: actor-network theory and rhetorical practice in scientific controversies" *Southern Communication Journal*, 76: 120–136.

Blok A. (2011) "Clash of the eco-sciences: carbon marketization, environmental NGOs and performativity as politics" *Economy and Society*, 40: 451–476.

Bond P. (2012) *The CDM in Africa cannot deliver the money* University of KwaZulu-Natal, Durban.

Bond P. and Dorsey M.K. (2011) "Anatomies of environmental knowledge and resistance: diverse climate justice movements and waning eco-neoliberalism" *Journal of Australian Political Economy*, 66: 286–316.

Both Ends (2016) Barro Blanco 1–5 (www.bothends.org/en/News/Our-stories-oud/Barro-Blanco/) Accessed 10 October 2016.

Bracking S. (2014) "The anti-politics of climate finance: the creation and performativity of the Green Climate Fund" *Antipode*, 47: 281–302.

Bracking S. (2015a) "Performativity in the green economy: how far does climate finance create a fictive economy?" *Third World Quarterly*, 36: 2338–2357.

Bracking S. (2015b) "The anti-politics of the Green Climate Fund: what is left to negotiate?" *Environmental Justice Organisations, Liabilities and Trade*, 34–41.

Bracking S. and Dorsey M.K. (2015) "Escaping the 20th century neoliberal prison complex: what should climate justice organizations do with the Green Climate Fund in the 21st century?" in Gabello J. and Gilbertson T. (eds) *Paths beyond Paris: movements, action and solidarity towards climate justice* Carbon Trade Watch, Paris, 47–50.

Brand U. (2010) "Sustainable development and ecological modernization – the limits to a hegemonic policy knowledge" *Innovation: European Journal of Social Science Research*, 23: 135–152.

Brown L., Polycarp C. and Spearman M. (2013) *Within reach: strengthening country ownership and accountability in accessing climate finance*, World Resources Institute, Washington, DC.

Brundtland, G.H. et al. (1987) Report of the World Commission on Environment and Development: Our Common Future, Oslo.

Bruun J.A. (2016) "Redirecting the global flow of money" *Foresight*, 1: 62–63.

Bruun J.A. (2017) Governing climate finance: paradigms, participation and power in the Green Climate Fund. Unpublished PhD thesis, University of Manchester.

Bumpus A.G. and Liverman D.M. (2008) "Accumulation by decarbonization and the governance of carbon offsets" *Economic Geography*, 84: 127–155.

Campbell L. and Brosius J.P. (2010) "Collaborative event ethnography: conservation and development trade-offs at the Fourth World Conservation Congress" *Conservation and Society*, 8: 245–255.

CAN (2011) *CAN-International: Comments on "Draft report of the Transitional Committee" of 7 October submission to: Transitional Committee of the Green Climate Fund* Climate Action Network, Washington, DC.

CAN (2013) *CAN-International's response to questions related to the "Paradigm Shift" posed in the co-chair's information note on the informal discussion on the business model framework of the Green Climate Fund* Climate Action Network, Washington, DC.

Caravani, A. et al. (2012) *The global climate finance architecture* Henrich Böll Stiftung, Washington, DC.

Carbon Market Watch (2014a) An 'exclusion list' for the Green Climate Fund (http://carbonmarketwatch.org/an-exclusion-list-for-the-green-climate-fund-wt/) Accessed 10 November 2015.

Carbon Market Watch (2014b) *Santa Rita: CDM hydro dam in Guatemala* Carbon Market Watch, Brussels.

Carr D.L. and Norman E.S. (2008) "Global civil society? The Johannesburg World Summit on Sustainable Development" *Geoforum*, 39: 358–371.

CDM (2012) *Climate change, carbon markets and the CDM: a call to action* CDM Policy Dialogue, Luxembourg.

Corbera E. and Brown K. (2010) "Offsetting benefits? Analyzing access to forest carbon" *Environment and Planning A*, 42: 1739–1761.

Dean M. (2010) *Governmentality, power and rule in modern society* Sage, London.

Death C. (2011) "Summit theatre: exemplary governmentality and environmental diplomacy in Johannesburg and Copenhagen" *Environmental Politics*, 20: 1–19.

Edwards M. (2008) "Have NGOs 'made a difference?' From Manchester to Birmingham with an elephant in the room" in Bebbington A.J., Hickey S. and Mitlin D.C. (eds) *Can NGOs make a difference? The challenge of development alternatives* Zed Books, London, 38–55.

Fisher W.F. (1997) "Doing good? The politics and antipolitics of NGO practices" *Annual Review of Anthropology*, 26: 439–464.

Flyvbjerg B. (2005) "Design by deception: the politics of megaproject approval" *Harvard Design Magazine*, 50–59.

Foucault M. (1991) "Governmentality" in Burchell G., Gordon C. and Miller P. (eds) *The Foucault effect: studies in governmentality* Harvester Wheatsheaf, London, 87–104.

Foucault M. (1998) *The will to knowledge: the history of sexuality Vol. 1* Hurley R. ed. Penguin, London.

Franks D.M. et al. (2014) "Conflict translates environmental and social risk into business costs" *Proceedings of the National Academy of Sciences of the United States of America*, 111: 7576–7581.

GAIA (2016) Seeking action against questionable CDM project Okhla Waste to Energy Plant, Delhi, India. *Petition* (https://www.change.org/p/cdm-executive-board-unfccc-seeking-action-against-questionable-cdm-project-okhla-waste-to-energy-plant-delhi-india?recruiter=116587535&utm_source=share_petition&utm_medium=facebook&utm_campaign=share_for_starters_page&utm_term=des) Accessed 1 September 2016.

Gane M. (2008) "Foucault on governmentality and liberalism" *Theory, Culture & Society*, 25: 353–363.

GCF (2011) *Governing instrument for the Green Climate Fund* The Green Climate Fund, Songdo.

GCF (2013a) *Business model framework: objectives, results and performance indicators* Songdo The Green Climate Fund, Paris.

GCF (2013b) *Decisions of the board – Fifth Meeting of the Board* The Green Climate Fund, Paris.

GCF (2014a) *Initial results management framework of the fund (progress report)* The Green Climate Fund, Bali.

GCF (2014b) *Investment framework* The Green Climate Fund, Songdo.

GCF (2015a) *Analysis of the expected role and impact of the Green Climate Fund* The Green Climate Fund, Songdo.

GCF (2015b) *Mandate and governance* The Green Climate Fund, Songdo.

GCF (2015c) *Current consultant list onsite consultant* The Green Climate Fund, Songdo.

Hadden J. (2015) *Networks in contention: The divisive politics of climate change* Cambridge University Press, New York.

Hajer M.A. (1995) *The politics of environmental discourse: ecological modernization and the policy process* Oxford University Press, Oxford.

Hajer M.A. (1997) "The politics of environmental discourse: ecological modernization and the policy process" *Computers in Human Behavior*, 17: 344.

Harmeling S. et al. (2013) *How can the Green Climate Fund initiate a paradigm shift?* Climate and Development Knowledge Network, London.

Huber A. and Joshi D. (2015) "Hydropower, anti-politics, and the opening of new political spaces in the Eastern Himalayas" *World Development*, 76: 13–25.

Hulme D. (2008) "Reflection on NGOs and development: the elephant, the dinosaur, several tigers but no owl" in Bebbington A.J., Hickey S. and Mitlin D.C. (eds) *Can NGOs make a difference? The challenge of development alternatives* Zed Books, London, 337–346.

IFC (2016) Summary of investment information (http://ifcext.ifc.org/ifcext/spiwebsite1.nsf/ProjectDisplay/SII31458) Accessed 1 March 2017.

Igoe J. and Kelsall T. (eds) (2005) *Between a rock and a hard place: African NGOs, donors and the state* Academic Press, Durham, NC.

IPCC (2013) *Working Group I contribution to the IPCC Fifth Assessment Report Climate Change 2013: the physical science basis summary for policymakers* Intergovernmental Panel Climate Change, Bonn.

Jasanoff S. (1997) "NGOs and the environment: from knowledge to action and the environment: from knowledge to action NGOs" *Third World Quaterly*, 18: 579–594.

Jasanoff S. (2013) "A world of experts: science and global environmental constitutionalism" *Boston College Environmental Affairs Law Review*, 40: 439–452.

Latour B. (1999) *Pandora's hope: essays on the reality of science studies* Harvard University Press, Cambridge, MA.

Li T.M. (2007) *The will to improve: governmentality, development, and the practice of politics* Duke University Press, Durham, NC.

Litfin K. (1994) *Ozone discourses. Science and politics in global environmental cooperation* Columbia University Press, New York.

Lohmann L. (2012) "Beyond patzers and clients" *Development Dialogue*, 61.

Lousley C. (2015) "Narrating a global future: our common future and the Public Hearing of the World Commission on Environment and Development" in DeLoughrey E., Didur J. and Carrigan A. (eds) *Global ecologies and the environmental humanities: postcolonial approaches* Routledge, New York, 245–267.

Lövbrand E. and Stripple J. (2011) "Making climate change governable: accounting for carbon as sinks, credits and personal budgets" *Critical Policy Studies*, 5: 187–200.

Lovell H. and Mackenzie D. (2011) "Accounting for carbon: the role of accounting professional organisations in governing climate change" *Antipode*, 43: 704–730.

Luke T. (1999) "Environmentality as green governmentality" in Darier E. (ed.) *Discourses of the environment* Blackwell, Oxford, 121–151.

MacKenzie D. (2009) "Making things the same: gases, emission rights and the politics of carbon markets" *Accounting, Organizations and Society*, 34: 440–455.

Methmann C.P. (2010) "'Climate Protection'" as empty signifier: a discourse theoretical perspective on climate mainstreaming in world politics" *Millennium – Journal of International Studies*, 39: 345–372.

Methmann C.P. (2011) "The sky is the limit: global warming as global governmentality" *European Journal of International Relations*, 19: 69–91.

Methmann C.P. (2014) *We are all green now: hegemony, governmentality and fantasy in the global climate polity,* Universität Hamburg Fakultät, Hamburg.

Müller B. (2013) "Same old, same old ... Too late for a paradigm shift? It must be said!" *Open letter to the GCF Board and supporters of enhanced/devolved access in the wake of the Paris Board meeting (GCF/B.05)*, Oxford Climate Policy, Oxford, available at (http://www.oxfordclimatepolicy.org/publications/documents/Too_late_GCF_blog.pdf) Accessed 20 September 2018.

Munnings C., Leard B. and Bento A. (2016) *The net emissions impact of HFC-23 Offset Projects from the Clean Development Mechanism* RFF, Washington, DC.

Nacpil L. et al. (2015) *CSOs letter to the board on accreditation* Songdo.

Nasiritousi N., Hjerpe M. and Linnér B.O. (2014) "The roles of non-state actors in climate change governance: understanding agency through governance profiles" *International Environmental Agreements: Politics, Law and Economics*, 14[3]: 265-280.

Nielsen T.D. (2014) "The role of discourses in governing forests to combat climate change" *International Environmental Agreements: Politics, Law and Economics*, 1–16.

Oels A. (2005) "Rendering climate change governable: from biopower to advanced liberal government?" *Journal of Environmental Policy & Planning*, 7: 185–207.

Oels A. (2013) "Rendering climate change governable by risk: from probability to contingency" *Geoforum*, 45: 17–29.

Olsen K.H. (2007) "The Clean Development Mechanism's contribution to sustainable development: a review of the literature" *Climatic Change*, 84: 59–73.

Orenstein K. (2016a) "Getting the GCF we fought for" *Friends of the Earth*, 1–4 (www.foe.org/news/blog/2016-07-getting-the--we-fought-for) Accessed 8 July 2016.

Orenstein K. (2016b) *Talking points to rebut arguments that the GCF needs to accredit big banks like HSBC, Credit Agricole, and Deutsche Bank*. Friends of the Earth, Washington, DC.

Oxfam et al. (2015) *Ten reasons why climate initiatives should not include large hydropower projects* Oxfam, London.

Pallas C.L. and Urpelainen J. (2012) "NGO monitoring and the legitimacy of international cooperation: a strategic analysis" *Review of International Organizations*, 7: 1–32.

Pollard A. and Court J. (2008) "How civil society organizations use evidence to influence policy processes" in Bebbington A.J., Hickey S. and Mitlin D.C. (eds) *Can NGOs make a difference? The challenge of development alternatives* Zed Books, London, 133–153.

Purdon M. (2013) "Neoclassical realism and international climate change politics: moral imperative and political constraint in international climate finance" *Journal of International Relations and Development*, 20(2): 263–300.

Raman M. (2013a) Board agrees on initial result areas and indicators *TWN Info Service on Climate Change.*

Raman M. (2013b) Disagreement over mobilising resources *TWN Info Service on Climate Change.*

Raman M. (2013c) GCF Board debates objectives, results and indicators *TWN Info Service on Climate Change.*

Redman J. and Orenstein K. (2015) *Getting climate finance right: successful examples for the Green Climate Fund from around the world* Washington, DC.

Rowling M. (2014) Green Climate Fund to back energy "paradigm shift" *Reuters* (www.reuters.com/article/us-climatechange-energy-idUSKBN0JO14620141210) Accessed 10 January 2016.

Schalatek L. (2011) *"Sub-optimal" outcome in the Transnational Committee: No consensus on the design of the Green Climate Fund* Henrich Böll Stiftung, Washington, DC.

Schalatek L., Nakhooda S. and Bird N. (2011) *Climate finance fundamentals: the design of the green* Henrich Böll Stiftung, Washington, DC.

Schalatek L. and Nakhooda S. (2012) *The Green Climate Fund* Henrich Böll Stiftung, Washington, DC.

Schalatek L. and Nakhooda S. (2013) *The Green Climate Fund: Climate finance fundamentals 11* Henrich Böll Stiftung Washington, DC.

Schalatek L. and Nakhooda S. (2014) *Info: the Green Climate Fund.* Henrich Böll Stiftung, Washington D.C.

Schalatek L. et al. (2015) *Green Climate Fund and the Private Sector* Henrich Böll Stiftung, Washington, DC.

Speth J.G. and Haas P.M. (2006) *Global environmental governance* Island Press, Washington, DC.

Stripple J. and Bulkeley H. (2014) *Governing the climate* Cambridge University Press, Cambridge.

Sullivan S. (2013) "Banking nature? The spectacular financialisation of environmental conservation" *Antipode*, 45: 198–217.

Sullivan S. (2016) "What's ontology got to do with it? On the knowledge of nature and the nature of knowledge in environmental anthropology" *in* Kopnina H. and Shoreman-Ouimet E. (eds) *Routledge international handbook of environmental anthropology* Routledge, London, 155–169.

The Adaptation Fund (2016) About the Adaptation Fund (https://www.adaptation-fund.org/about) Accessed 1 February 2016.

UNECE (1998) *Convention on access to information, public participation in decision-making and access to justice in environmental matters (The Aarhus Convention)* United Nations Economic Commission for Europe, Aarhus.

UNECE (2005) *Almaty Guidelines* United Nations Economic Commission for Europe, Almaty.

Uvin P., Jain P.S. and Brown L.D. (2000) "Think large and act small: toward a new paradigm for NGO scaling up" *World Development*, 28: 1409–1419.

Vertis (2016) *Carbon prices 2016* Vertis Environmental Finance, Budapest.

Watt R.M. (2016) The moral economy of carbon offsetting: ethics, power and the search for legitimacy in a new market. Unpublished PhD thesis, University of Manchester.

Winkler H. and Beaumont J. (2010) "Fair and effective multilateralism in the post-Copenhagen climate negotiations" *Climate Policy*, 10: 638–654.

Wu B. (2013) Fighting for the soul of climate finance *Huffington Post* (www.huffingtonpost.com/brandon-wu/fighting-for-the-soul-of-_b_4250874.html).

Wu B. (2015) *Strategic plan for the Green Climate Fund: inputs from ActionAid* Action Aid, Washington, DC.

10 Water values and the negotiation of water use

Philip Woodhouse and Mike Muller

Introduction: a contested vision of water valuation

The question of water valuation has split the water sector for more than a decade, following the World Water Vision (WWV) presented during the World Water Forum (WWF) in the Hague in 2000 (WWC 2000). In particular, the WWV emphasis on the importance of 'full-cost pricing' as a necessary condition to attract private-sector investment to undertake the 'business of providing water services' provoked a hostile reaction from many quarters. NGOs especially criticised it for not respecting the 'rights, knowledge and experience of local people and communities' (Morley 2002, 7).

Arguments about the role of water pricing and water markets continue, but are today more attenuated by the widespread recognition that 'price' does not, or should not, be equated with 'value'. In this chapter we review the question of value as it applies to water and examine approaches to how a diversity of water values can be reconciled within decision-making processes. We identify two separate issues to be addressed. The first is to characterise water as a complex resource with a diversity of uses, broadly defined, and to consider the values that may be associated with water and why they are important. The second is to consider how societies can, and do, make decisions about water resources where different values systems intersect and overlap. The ability of social decision-making processes to reconcile different values systems is critical to mediating competing demands for water resources. It is therefore important to identify generic principles that can guide the development of appropriate decision-making approaches and their application.

We argue that the approaches that have guided policy and practice over the past three decades have suffered from a reliance on reducing all water uses to commensurability consistent with a single basis of value, either economic or ecological. We outline an analytical values-based approach that seeks to identify and recognise more explicitly a plurality of values in water resources and their use. We argue that such an approach can elucidate the roles of the state and other sectors of society in water management and, further, may help to determine a more clearly defined, if limited, role for markets and prices, and in what contexts they may be useful. We start by considering why the question of economic valuation and commoditisation of water has proved so divisive in principle and problematic in practice. We then examine the rise and demise of 'integrated' water management based on hydrological boundaries defined as river basins.

An uncooperative commodity?

In this section we trace the debate on economic valuation of water and its linkage to efforts to commoditise water in order to achieve more efficient utilisation. We highlight how, despite economic valuation being a widely accepted principle of water policy, its application in practice has been highly conditioned by the specific patterns of political power relations in local contexts. This raises questions not only about the commensurability of values applied to different uses of water, but also to the primacy of economic valuation as a framework for understanding political decisions about water.

The commonplace beginning for any such discussion is the observation that water has a number of characteristics that make allocation through markets complex (Dellapenna 2012; Woodhouse and Muller 2017). Water takes a multiplicity of forms (solid, liquid and vapour) and qualities (principally freshwater and saline sea water). It is renewable, in many cases through an annual seasonal cycle, although in others (notably in the case of underground water), over cycles of decades or centuries. These different temporalities have implications for what are considered water 'stocks' and water 'flows'. It may be a fugitive resource (water temporarily in a stream through which it flows) or can be considered to be a common pool resource (such as groundwater, a stock that can be accessed by a number of users). It can also be re-useable.

The earth's 'stock' of freshwater is only 2.5 per cent of the total (the balance is salt-water in oceans). Most of this freshwater (75 per cent) is 'locked' in glaciers and icecaps. Much of the rest (a further 22 per cent) lies relatively inactively, underground. Only 0.53 per cent is in the 'visible' water cycle, in lakes and rivers, in the atmosphere, as clouds and rain as well as in the top layers of soil where it supports plant life (Agnew and Woodhouse 2011, 3). Estimates (e.g. FAO Aquastat 2017) put human use of this 'renewable freshwater' at about 10 per cent when aggregated across the globe. Human usage of freshwater 'flows' is sometimes conceptually distinguished between 'blue' water, the visible mobile water in rivers and lakes, and 'green' water, the rainfall cycle that infiltrates into the ground and then evaporates from the soil surface or via plants without becoming 'blue water'. Smaller streams of water that have been degraded by use are sometimes described as 'grey water' (general domestic and urban wastewaters, with limited pathogenic content and relatively easily recycled) and 'black water' (sewage effluent containing pathogenic human and animal wastes).

The capacity for water to serve multiple uses and to be re-used has also given rise to a terminology of 'consumptive' use (e.g. irrigation or cooling by evaporation) by which water is made unavailable to other users, and 'non-consumptive' use (cleaning, cooling, waste disposal, power generation, navigation) whereby water is withdrawn and then returned to stocks of 'blue' water (rivers, lakes, etc.), or not withdrawn at all (ecosystems services, fishing and aesthetic, cultural and recreational services). Non-consumptive uses may nonetheless require water to be available in a particular location at a particular time with a particular depth, flow or purity – thus opening scope for conflicting requirements of different uses.

We may see, therefore, that water is a renewable resource that, on the basis of the figures given above, is not scarce at a global level. However, it is also clear that water will become scarce in places where demands are multiplying and intensifying. In practice, this means that as local water use increases in a particular location, greater organisation and investment of labour and capital is needed to access additional sources of water. The growth of urban centres during the 19th and 20th centuries was often accompanied by the capture of ever more distant water resources using water infrastructure on an increasing scale. This was accompanied by a stronger social control of water supplies, typified in the takeover of private water companies by municipal governments in the late 19th and early 20th centuries and by increasing use of state funding for the construction of dams, aqueducts, sewers, water treatment facilities and so on (Swyngedouw 2006). Until the last quarter of the 20th century the model of state funding for water infrastructure was largely uncontested and was applied on an ever-greater scale. In the US alone, some 54,000 dams were constructed between 1930 and 1979, of which at least 1,700 were more than 30m high (Reisner 1993; NID 2016). As well as controlling the funding of water infrastructure, the state also assumed powers in relation to the planning of and decision-making around water use. Criticism of the quality of this decision-making from two quarters (those who questioned the social and political effects and those who opposed its financial costs) steadily undermined the legitimacy of government agencies. By the 1980s, their control of investment and management was increasingly threatened by policies aimed at reducing public expenditure and transferring water management to commercial enterprises.

Important changes in the discourse of water management accompanied this shift in control. Whereas rising demand for water had previously been met by increasing (state) investment in infrastructure, the shift to commercial management was accompanied by a reorientation in management goals towards 'demand management' (i.e. to reduce water consumption). Ostensibly, this was to allow a reduction in costs, at least in those countries where large infrastructural investment had already been made. A central principle in reducing water use was the pricing of water to encourage more efficient use. The narrative of increasing efficiency promoted by commercial water management was most prominent in relation to 'water services' – the delivery of treated water and disposal of wastewater – for businesses and domestic households. However, the wider principle of price as a means of allocating water between competing demands was extended to the management of the broader water resource constituted by all forms of 'blue' water (lakes, rivers, groundwater aquifers). Underlying the use of price, and the economic valuation techniques through which it was to be revealed, lay a narrative of water as a scarce resource that could not satisfy all the competing and growing demands upon it. We shall return to the narrative of water scarcity below, but first we wish to highlight the difficulties involved in pricing water.

Karen Bakker's early critique of the divestiture of UK state water management to commercial companies (Bakker 2000, 2003) branded water "an uncooperative commodity", suggesting the materiality of water makes it inherently difficult to commoditise. As we discuss below, subsequent writing has deviated from that interpretation. Yet, the narrative of water as a tradeable good allocated through price was always problematic. Many of the uses identified above, for example, are not commensurable in volumetric terms. The use of water for irrigation, industrial cooling, domestic (cooking and hygiene) use and hydropower may all be quantified in terms of volumes of water needed to achieve particular outcomes. Other uses may require a stock (minimum depth: e.g. navigation), while yet others demand flow characteristics (height difference: e.g. hydropower, some fisheries). Other dimensions that may be at least as critical as volume are: timing (e.g. dry-season irrigation), and reliability (assurance of supply: supplies to domestic and industrial users). The dimensions of volume and risk will vary in significance between different types of use. This suggests that the logic of a single unit price of water does not apply across all uses of water. Even more fundamentally, there are uses of water that are not only difficult to quantify in volume terms, but also to envisage at the level of the individual user. Aesthetic, cultural and ecological dimensions of water resources are notoriously hard to capture in any but broad social terms, and water is often claimed as a national resource, for which the state is ultimately responsible.

Although this sense that water is 'irreducibly plural' has prompted scepticism about the scope for water allocation via 'pure' markets, it is the political dimensions of 'trade-offs' between different water uses that have made the question of economic valuation of water acute (Dellapenna 2012). Pipelines abstracting water for use in distant towns at the expense of irrigated agriculture and rural livelihoods, or upstream irrigators diverting water from rivers feeding hydropower turbines raise questions that are not fully addressed by the mechanics of cost–benefit analysis (CBA). These questions are manifest even in cases where the 'market' logic of water allocation seems inescapable. Thus, the notorious case of the diversion of the Owens River into the Los Angeles aqueduct at the start of the 20th century is often cited as an example of urban water users outbidding agriculture in terms of the price they could pay for water. Yet, in practice 'payment' for water rights was achieved by covert land purchases on behalf of the LA Department of Water and Power (LADWP) because they would have been unaffordable on an 'open' market (Reisner 1993). Moreover, the financial logic of diverting the Owens River in the 1920s doubtless did not take account of the USD 1 billion spent by the LADWP in the first decade of the 21st century in attempts to control the toxic dust storms generated by the salt flats created where Owens Lake – fed by the Owens River – used to be.[1]

Contemporary contestation over water allocation repeatedly demonstrates that the economic valuation of water is subordinated by context-specific political factors. Thus, while industrial use of the Orange River, which supports approximately 60 per cent of South Africa's economy, is largely unchallenged by consideration of the value of the ecosystem services foregone, the Okavango, a river of a similar size, is managed entirely to satisfy ecological goals. The Permanent Okavango River Basin Water Commission (OKACOM) has used CBA to demonstrate that the value derived by the use of water for upstream economic and social purposes, particularly for irrigation in Namibia, would be outweighed by the losses suffered in "ecosystems goods and services" in the Ramsar-recognised[2] wetland downstream in Botswana that is the centre of a significant eco-tourism industry (Aylward 2009). In this case the high visibility of the Okavango wetland to the international environmental movement significantly strengthens the ability of downstream interests to block upstream water use. The value of ecological services has been similarly argued to justify prohibition of irrigation by African farmers upstream of the Kruger National Park (Woodhouse 1995).

In other contexts, however, the political significance of small-scale producers may override the logic of CBA. In South Africa's Komati river basin, although the bulk of water is used by irrigated agriculture, constraints are applied to maintain a flow downstream to the estuary in Mozambique where a shrimp industry provides livelihoods for about 3,000 artisanal and semi-industrial fishers. The economic valuation of the shrimping activity, estimated to be USD3.5 million in the early 1990s (Monteiro and Matthews 2003), is large in terms of the Mozambican economy but small in relation to the value of using additional water for industrial expansion in South Africa. Yet, arguments that associated the health of the shrimping activity with the maintenance of biodiversity of the estuarine ecology ensured that a minimum cross-border flow was guaranteed as part of the river basin management. A particularly high-profile instance of the emblematic politics of 'smallholder livelihoods' resulted in the CocaCola company in India closing its bottling plant in the village of Plachimada in Kerala state, in 2000. Following allegations that the plant's extraction of groundwater was reducing supplies to local farmers, a campaign to close down the plant resulted in local government revoking the company's licence to extract water. The privileging of 'smallholders' needs to be seen in a context where the over-extraction of groundwater is widespread and well known in India, and has proved to be difficult to resolve, not least because a primary cause is the ability of farmers to pump from ever-deeper wells using state-subsidised electricity. The CocaCola dispute thus reflects the political clout of a farmers' lobby and the inability of government to regulate their groundwater abstraction (Shah 2010).

The significance of context is reinforced by evidence that social priorities in water management have changed over time. This is evident particularly in countries in which industrialisation has taken place. Since the turn of the century, increasing interest in the amenity and recreational use of rivers like the Rhine has paralleled changing policies towards water quality and the hydrological and ecological functioning of floodplains (Hooijer et al. 2004) that challenge the once-dominant primacy of navigation, hydropower and waste disposal in river management. In Korea, urban environmental renewal in Seoul has included the reinstatement of the Cheonggyecheon stream that had become a sewer with an urban motorway built over it. As an emblematic element of a politics of urban renewal, the process was reversed: the road was removed and a reconstructed river restored as the centrepiece of a new urban area (*The Guardian* 2006). An even more radical move is the proposal in parts of the US that dams built for hydropower and related purposes should be removed to 're-naturalise' the rivers. On the west coast, this campaign has focused on restoring the salmon runs and other fisheries that once characterised the rivers. On the east coast, the focus has been on removing dams that are no longer required. Such moves are not uncontested, though not necessarily from utilitarian perspectives such as hydropower generation. Only about 3 per cent of all dams produce hydropower. Dam removal threatens the value of properties developed alongside reservoirs, but non-monetary factors may actually be more important, if no less conflictual. Some value the aesthetics of water tumbling down from dams or the recreation value of reservoirs, while others seek to preserve industrial history (Scientific American 2013).

The discussion above suggests that while economic valuation has been a source of major contestation in discussions of water management, its application in practice appears heavily influenced by the specific context of political power relations. This raises the questions of how particular valuation systems have dominated academic and policy debates and why they appear to enjoy rather less supremacy in practice.

Framing discourses of water governance

Decisions on how to reconcile different, and potentially competing, water uses have for the past 20 years been predominantly guided by the idea of integrated water resource management. These have been based, above all, on the 'Dublin Principles' of Integrated Water Resource Management (IWRM), originally negotiated at the International Conference on Water and the Environment (ICWE) in Dublin in 1992 (Young et al. 1994). Three key principles emerged. First, water should be managed within natural hydrological units constituted by river basins or watersheds within which water was identified as a renewable, but finite (and hence potentially scarce) resource. Second, since water was scarce, it had to be treated as an economic good. Third, to resolve inevitable competition over access to this scarce resource, participative approaches had to be adopted within the boundaries of the water course.

At one level, it may be argued that these three principles represent an attempt to reconcile three quite distinct, and possibly contradictory, perspectives on water management arising from hydrological engineering, economics and notions of deliberative democracy (Habermas 1962; Dryzek 2000). However, it is also evident that the three principles are united by a discourse of water scarcity that Mehta (2010, 2) calls a 'totalising discourse', obscuring context-specific political questions of water allocation in favour of universal normative technical criteria. The water scarcity narrative supports two important ideological positions. First, it demands efficiency in water allocation and recourse to economic valuation of implicitly competing uses of water. It is axiomatic that for water to be treated as an economic good, it must be defined as scarce. Second, a narrative of water scarcity emphasises water conservation and, within industrialised societies, benchmarks of water management that minimise departure from 'natural hydrology' as the basis of 'sustainable' water resource management.

As observed above, the aims of returning rivers to more natural conditions are often to rectify past 'mistakes' in water management: increased flood risks due to channelling of river beds and encroachment of human settlement and industry on floodplains; reduced fish and other aquatic biodiversity due to pollution and obstacles to fish migrations to spawning sites. The redress of such 'excessive' development is thus framed as returning rivers to more natural conditions. This, in turn, can be construed as a commitment to manage water as far as possible to maintain its natural hydrology, with attendant assumptions that the less water use (i.e. withdrawal for social and economic activity) that takes place the better. While such management goals have emerged in many industrialised economies during the past 40 years or so, their extension to the Global South is evidently inconsistent with the need in some regions (e.g. sub-Saharan Africa) to use more water as part of increasing economic activity to raise living standards. From a 'development' perspective, therefore, using natural hydrological patterns and their associated ecologies as benchmarks for sustainability risks making water 'scarce' for all other goals.

Luks (2010) has argued for a distinction between 'limits' as objective, empirically verifiable characteristics, and 'scarcity' as individual or social subjective perceptions of what those limits signify. For example, water limits may be measured in terms of flow volumes in streams or recharge rates of aquifers. However, what this signifies in terms of water use for particular purposes is subject to cultural and political values and priorities. Luks goes further, to argue that the use of ecological models to construct 'scarcity' is as much a social construction as economists' assertion of competing uses of water. This is underlined by his (Luks 2010, 99) comparison of 'ecological' and 'environmental' economics: "finiteness, limits and 'absolute scarcity' are just as much social constructions as markets, prices and 'relative scarcity'". From the perspective of water management, the assertion (as in ICWE interpretations of IWRM) of water basins as natural units within which water allocation decisions must be resolved transforms one perception – the interconnectedness of different uses of water – into another – an impermeable boundary within which all water uses must compete. The pervading significance of the underlying narrative of scarcity in justifying water pricing is evident even in the EU water framework directive: "not putting a price on a scarce resource like water can be regarded as an environmentally-harmful subsidy" (EU 2012, 10). Yet in much of temperate Europe water is not subject to shortage. In this sense, therefore, scarcity has an important performative role in establishing the norms by which the value of different uses of water is assessed.

'Calculative devices' for assessing water value

The 'scarcity' narrative framing water generates 'universal' criteria for management. Within the ICWE these were identified as: management according to hydrological units; treatment as an economic good; and allocation by institutions representative of 'stakeholders' in the water resource. From these normative procedures follow a set of techniques for determining the adequacy or otherwise of water management in meeting goals: the preservation of minimum flow rates in rivers; the sustaining of particular ecologies or populations of selected species; or the generation of economic output or jobs per unit of water. These non-political 'technical' criteria for good management (cf. Ferguson 1994; Li 2007) have the effect of obscuring context-specific political questions about water allocation.

First, watershed boundaries emphasise 'natural' limits to supply, but thus foreclose options of bringing water from neighbouring areas and instead focus on managing competing demand for a 'scarce' resource. Then, management of competing demands for water from different users is to be achieved by either pricing water through a market and/or assessing the economic value of different uses of water and creating participatory institutions to decide how water should be allocated. In both cases, the 'neutrality' of such technical mechanisms to bring about fair distribution of water is an important legitimating factor. However, in practice, water markets suggest allocation on the basis of purchasing power. Even a more ostensibly political process using participatory institutions can be 'rendered technical' where the terms of discussion are determined by hydrological and economic parameters that circumscribe the scope for alternative decisions, and may be subject to a strongly asymmetric distribution of information among stakeholders that heavily favours existing large users of water and the approaches to valuation of water that they favour (Brown 2011, 2013; Peters and Woodhouse, Chapter 11 this volume).

These observations suggest that whereas the Dublin principles sought to reconcile different uses of water, they did so within a fixed frame of water values informed by a discourse of scarcity. We argue below that this is too restrictive to cover the range of values that apply to water in practice, and therefore a broader framework of water values is needed. Such a framework would need to encompass a more diverse range of water values, including at least some which may be incommensurable with others.

Institutional assemblages

Although, as we indicated earlier, technical approaches involving economic valuation of alternative water uses and CBA have in practice been subject to context-specific political economy, it has been the identification of such principles with the activities and viewpoints promoted by specific organisations and institutions comprising a 'water sector' (private companies, river basin authorities, professional associations, academic departments, etc.) that has generated widespread resistance.

While the 1992 ICWE in Dublin was convened by the 'Inter-secretariat Group on Water Resources' (representing 24 UN agencies, and attended by 500 participants from 114 countries, 28 UN agencies and 58 non-government organisations) (Young et al. 1994), the policy agenda with which the 'Dublin Principles' became associated was promoted by a 'World Water Council' (WWC) set up in 1995 with a headquarters in Marseille, to provide a 'multi-stakeholder platform' to encourage debates and exchange of experience in order to generate a "common strategic vision on water resources and water services management amongst all stakeholders in the water community" (WWC 2010).

The WWC embarked on a series of WWFs held every three years, the second of which, in The Hague in 2000, attracted some 5,700 participants, including 114 government ministers and an additional 32,500 visitors to a 'World Water Fair' associated with the forum. This fair saw the launch of the controversial WWV we referred to at the start of the chapter, a report by a 'World Commission on Water in the 21st Century' convened by the WWC. The WWV used the Dublin Principles to assert the primacy of technical principles: 'holistic' management of the water resource using catchments or basins as units of management, rather than 'artificial' political or administrative boundaries (WWC 2000, 25). While declaring 'participation is key' with community groups and user associations playing a role at a local scale and "empowering women's groups, the poor, youth, and community-based groups to have an adequate voice" in decision-making 'is a necessary pillar' of water management, it argued that water must be recognised as a scarce commodity, whose price should provide an incentive to transfer use from inefficient to 'higher-valued' uses. This theme was further developed as an argument that the 'gloomy arithmetic' of population growth and rising demand for urban and agricultural water was beyond the fund-raising capacity of governments and public-sector water utilities, and consequently necessitated allowing the private sector to get on with the 'business' of providing water services. The assertion that 'full cost pricing of water' was ' the single most immediate and important measure' (WWC 2000, 33) was interpreted by many as advocating the expansion of operations of large, mainly French- and British-based, water companies to take over the provision of water services in countries of the Global South.

This established the WWC and WWF as a highly visible institutional representation of the type of 'private sector' water concession that was proving highly conflictual. The Cochabamba concession in Bolivia, held by a consortium led by a company owned jointly by Bechtel from the US, and United Utilities from the UK, was cancelled by the Bolivian government the month after WWF in The Hague, following two weeks of rioting which left six people dead. Subsequent WWFs became increasingly targeted by opponents of the WWV 'business' model of water service who set up 'Alternative Water Forums' as platforms to accuse the WWF of being dominated by commercial interests. Much of the 'alternative' to the WWV was articulated in terms of 'human rights' to water taking precedence over economic valuation of water use. By the Fourth World Water Forum in Mexico City, the summary of the forum observed the right to water was "for the first time debated in a rather consensual spirit" (WWC 2006, 2) and that:

> There is a strong need to initiate dialogue within civil society on the involvement of the private sector to choose a model people want to follow for provision and enhancement of their water services. Water service provision is and will remain mainly public. ... Private sector provision must be considered as only one of the possibilities taking into account the accumulated knowledge and previous experiences.
>
> (WWC 2006, 52)

It concluded:

> Local authorities have an essential role in generating finance for investments in water services. ... This necessity is founded on the recognition that users and taxpayers are in the end the main financiers, and on the associated shift from full-cost recovery to a solidarity system of fair tariffs combined with targeted subsidies.
>
> (WWC 2006, 108)

The emergence of an alternative narrative encompassing non-economic and non-market criteria was thus associated with opposition to a particular set of actors – commercial water companies – whose control of decisions about water was perceived as illegitimate, and was marked by further cancellations of concessions held by multinational water companies, notably in Dar-es-Salaam (in 2005) and Buenos Aires (in 2007).

It is important to observe that the alternative narrative suggested above, while opposing the primacy of commercial logic in water allocation, emphasises that alternative approaches need to be context-specific, rather than following a single management model. This point is underlined by detailed analysis of the Cochabamba case, which suggests that portraying events as a simple confrontation between the urban poor and the multinational corporation would be highly misleading (cf. Nickson and Vargas 2002). Similarly, in the Indian context, Mollinga (2010) has argued that conflicts over water are conduits for more systemic polarisations within society. Nonetheless, the visibility of the multinational water companies may be argued to have contributed a clear target that made it more possible for opponents to attack the narrative of economic valuation as a basis for allocation of water. In the next section we consider the emergence of alternative narratives of water use and their relationship to values systems in water.

From 'rendering technical' to 'recognising political'

The discourse of a 'human right' to water played a major role in challenging the dominance of economic valuation of water in water policy, and has been adopted as an underlying rationale for international benchmarks for improving water services, as in the UN Millennium Development Goals and their successors since 2015, the Sustainable Development Goals. A human right to water has also been recognised in some practical measures, such as the South African government's guarantee of 'free basic water' of 6,000 litres per household per day. However, the utility of this narrative as a basis for governing water allocation and improving water services has long been questioned, even by critics of water privatisation (cf. Bakker, 2008). This is consistent with a more general move away from 'universal' narratives of water governance towards one where water is but one dimension of deliberated processes that are context-specific. It is arguable that this narrative shift is evident in two quite distinct arenas of debate, one more firmly anchored in political economy/political ecology and the other emerging from within a 'water sector' literature that engages, at least in part, debates among water 'practitioners'.

The first of these has been concerned with a critical reassessment of the notion of water as an 'uncooperative' commodity. Walker's (2014a, 2014b) study of the privatised water industry in the UK revisited the failure to establish a market in water services (commoditisation) – contrasted with the successful transfer of ownership and management to private corporations – in the early 2000s. This failure had originally been attributed to difficulties of reconciling the flowing, fungible nature of water resources with the demands of a market for a static, tradeable property, and associated high entry (infrastructure) costs (Bakker 2003). Walker observed that the UK coalition government of 2010 re-worked the role of markets in water governance. Rather than 'emerging' through laissez-faire deregulation, markets were deemed to be a product of institutional investment. The state was thus seen to be responsible for delivering the efficiencies of market allocation. Specifically, 'Ofwat' the governmental Water Services Regulation Authority for England and Wales was to be responsible for ensuring competition between privatised water companies (a demand that was to be later made of other privatised utilities in the UK). In order to achieve this, water companies were to undertake investment in metering of domestic supply (to allow volumetric charging of water users, rather than the flat rate charges prevailing in the UK), and to engage in trade in abstraction licences, which were to be made of shorter duration to encourage innovation, and to integrate distribution networks to make transfers of water between companies easier. Walker explains the water companies' rejection of these proposals as a consequence of their impact on companies' financial risks. In particular, increased competition threatened a business model driven by very high assurance of supply to water users underpinned by investment in water supply infrastructure that relied on long-term licences and high capital value to generate high rates of return over time. Suggestions that companies' licences to abstract water would be reduced and reviewed more frequently to encourage water conservation were met by company threats of compensation claims.

Walker (2014b) argues that the UK water industry shows a failure of water commodification that is less determined by the intrinsic 'materiality' of water but also a consequence of the 'relational situated dimensions' of its materiality. This is consistent with Swyngedouw's (2004) assertion of the significance of historical and geographical insights into the 'hybridising' processes that influence the characteristics of material nature. This sense, that the capacity of water to circulate as a commodity is associated with particular historical and geographical (including social and political) circumstances is consistent with arguments made by Castree (2003) and Bakker and Bridge (2006). It is also congruent with O'Neill et al.'s (2008) interpretation of non-commensurability in the way different people value elements of nature. These are seen not just as 'resources' in a generic utilitarian sense but as entities that have specific significance arising from individual cultural and economic experience that constitute a 'narrative' of how value is understood for a particular element, such as a lake, stream or coastline.

In this narrative, water is not always 'uncooperative' to human projects of commoditisation due to its intrinsic unruliness, but is more or less cooperative according to the specific historical and geographical context. This specificity creates for different groups of people associations and values that are resistant to commensuration. This understanding converges with another critique of commensuration of water values in the literature on water management. This second critique focuses particularly on the interpretation of IWRM that emphasises the boundaries of watersheds or river basins as delimiting the units of hydrological management. While such hydrological units were once declared (e.g. by WWC 2000) as preferable to 'artificial' political administrative boundaries, later writing (Mollinga et al. 2007; Lenton and Muller 2009; Muller 2015) has argued that practical management of water to meet development needs is more usefully undertaken in multi-sector decision-making arenas or 'problemsheds' focused on social goals, rather than single-sector institutions focused on water conservation.

Associated with this multidimensional approach to water allocation is a retreat from 'universal' models of water governance. This is particularly striking where meta-analysis of a range of experiences has sought patterns of factors or their combinations that result in successful water governance, but have been able only to identify a list of 'scenarios' or 'syndromes'. The authors of one such study concluded that:

> A major challenge in the water sector has been how to effectively implement initiatives founded on a set of general principles ... In a heterogeneous world made up of different cultures, social norms, physical attributes, availability of renewable and non-renewable resources, investment funds, management capacities and institutional arrangements ... In the absence of a unifying framework, it has been difficult to diagnose what the objective of an IWRM intervention should be, which tools might be suitable for a specific region and whether implementation has been successful.
>
> (Srinivasan et al. 2012, 12)

Significantly, they concluded that while much policy is concerned with managing a 'demand–supply gap': "meta-analysis suggests that increased attention to the underlying sociotechnical drivers of demand and institutional structure will allow for more realistic scenario development" (Srinivasan et al. 2012, 12).

The retreat from universalising narratives that assume commensurability may be seen even in economic approaches to valuation. Thus Turner and colleagues argue for a contextual analysis that must: "encompass appropriate socio-economic, political, and cultural parameters. ... [T]o highlight the 'political economy' context and the need to recognize a plurality of values that may not be strictly commensurable" (Turner et al. 2010, 81). The same authors highlight the problems of many techniques used to determine individuals' economic valuation of natural phenomena, not least of which is that these are liable to change according to political and economic context.

If there is an emerging narrative that water allocation decisions need to be grounded in a plurality of values systems that will be encountered in each local context, what 'calculative devices' – benchmarks or norms – can be discerned as associated with such a narrative? In practice, while there is evidence of widespread interest in the practical necessity of understanding water values, not least among commercial users, procedures remain rooted in the pursuit of financial values based on welfare economics (e.g. WBCSD 2013). Nonetheless, such guidelines for good practice tend to be wider in scope than conventional environmental economic approaches would suggest, at least in acknowledging the incommensurable values as part of their reality. Thus the World Business Council for Sustainable Development (WBCSD) is at pains to emphasise that 'price' does not equate to 'value' and despite its welfare economics-based approach "advocates the recognition of and accounting for wider environmental, social and economic values outside of this conceptual framework" (WBCSD 2013, 8). In practice, this amounts to advocating for a qualitative approach as a first step in identifying the range of values systems relating to water held by people in a specific context, and the use of multi-criteria assessment techniques to try to assess the relative importance (value) that 'stakeholders' recognise for different dimensions of water. These statements suggest a move towards more deliberative approaches to negotiating incommensurable values in water. As does the following quote from Nestlé's 2012 'Shared Value Report', cited by the World Economic Forum (WEF) (2014):

> We believe that we can create value for our shareholders and society by doing business in ways that specifically help address global and local issues in the areas of nutrition, water and rural development. This is what we mean when we speak about Creating Shared Value (CSV). We proactively identify opportunities to link our core business activities to action on related social issues.
>
> (WEF 2014, 10)

For many, deliberative processes form a central element of democratic process through which people can exercise their sense of values as citizens rather than as consumers. Yet, deliberative processes formed part of the 'universal' principles of water governance set out by ICWE and implemented in the form of water users' representation and participation in river basin management organisations. It is not difficult to find evidence that 'participatory' processes in water governance can be used to legitimate strongly asymmetric power relations (Brown 2013; Evans and Pratchett 2013). While this suggests scope for doing deliberation better, it also raises the possibility that the pursuit of 'shared values' may need to focus on outcomes rather than on processes. However, the context-specific sensitivity of such an approach is underlined by Goff and Crow's (2014) critical evaluation of the practical significance of 'access to drinking water' indicators used by international agencies to measure progress towards 'water equity'. They argue that the selection of metrics that chart 'basic access' to water via public standpipes grossly underestimates the extent to which people's livelihoods are constrained by limited access to water. Focus on 'drinking water' they argue is a poor measure of water equity that is more accurately determined by the large number of economic and social activities made feasible by a piped water supply within the household.

Advocates of an approach to water governance rooted in deliberative democracy that focuses on the water values manifested by people at different scales of decision-making emphasise the role of government. Mollinga et al (2007) set out three 'structural features' which should govern institutional dimensions of water: the state is the central promoter of reform; information and capacity-building is key to transparent and accountable decisions; and implementation needs coalitions of 'empowered stakeholders'. A similar institutional assemblage is evident in WEF (2014) where the solution to business concerns about risks posed by insecurity of water supply is stated as: "it will usually be necessary for the state to take the lead since one of their roles is precisely to reflect, represent and mediate between different interests and value systems" (WEF 2014, 13).

One aspect that such formulations leave implicit is that different sets of stakeholders will be present at different scales of decision-making and within a process of reconciling large-scale 'national interest' priorities with more local-scale values, most emblematically portrayed by the displacement of local communities and livelihoods by large-scale water infrastructure: dams. It is also evident that water commoditisation through market exchange is not universally rejected. While economic valuation may be rejected as a way to substitute 'market' principles for political decisions about allocations for different water uses, as discussed earlier, it may present an efficient and transparent means to allocate water among people who all use the water for the same purpose and who are therefore likely to use a common perception of value for water, as appears to be consistent with the experience of water-saving during drought episodes in Australia's Murray–Darling basin (Connor and Kaczan 2013). As with other institutions, markets reflect power relations, and expectations of fairness and efficiency are likely to be realised only to the extent that the broader political economy supports that outcome. Indeed, a consequence of the comparative – and definitely not absolute – demise of economic- or conservation-based narratives that rendered water management a question of technical norms is to highlight the political nature of the state and the quality of democratic processes. In terms of fairness of outcome, water allocation instruments will only be as effective as the wider political system allows.

We started this chapter by identifying water as a resource whose allocation and use among diverse users is a fundamental requirement of societies. We have traced the efforts over the past three decades to introduce universal principles that establish water allocation as subject to technical criteria that draw legitimacy from economic valuation and environmental economics but also from conservationist perspectives on environmental management. We have argued that such universalising technical approaches tend, in practice, to be heavily conditioned by local political relationships that expose the incommensurability of systems of value encountered among those who use water. Resolving competition between different water users thus becomes more a matter of explicitly political processes rather than one of application of technical economic or hydrological rules. This finding does not constitute a 'solution' but serves to highlight the importance of the political processes and values that underpin social decision-making.

Notes

1 http://digital.library.ucla.edu/aqueduct/owens-valley-environmental-hazard
2 The Convention on Wetlands, called the Ramsar Convention, is an intergovernmental treaty that provides the framework for national action and international cooperation for the conservation and wise use of wetlands and their resources.

References

Agnew C. and Woodhouse P. (2011) *Water resources and development* Routledge, Abingdon.

Aylward B. (2009) *Economic valuation of basin resources* Final Report to EPSMO Project of the UN Food & Agriculture Organization as an Input to the Okavango River Basin Transboundary Diagnostic Analysis: Final Draft, OKACOM, Botswana www.okacom.org/site-documents/tda-background-reports/economic-valuation-of-basin-resources-final-report-to-epsmo-project-of-the-un-food-and-agriculture-organization-as-an-input-to-the-okavango-river-basin-transboundary-diagnostic-analysis-final-draft/view (Accessed 9 February 2018).

Bakker K.J. (2000) "Privatising water, producing scarcity: the Yorkshire drought of 1995" *Economic Geography* 76: 4–27.

Bakker K.J. (2003) *An uncooperative commodity: privatising water in England and Wales* Oxford University Press, New York.

Bakker K.J. (2008) "The 'Commons' versus the 'Commodity': alter-globalisation, anti-privatisation, and the human right to water in the Global South" in Mansfield B. (ed.) *Privatization* Blackwell, Oxford, 38–63.

Bakker K. and Bridge G. (2006) "Material worlds? Resource geographies and 'the matter of nature'" *Progress in Human Geography* 30: 5–27.

Brown J. (2011) "Assuming too much? Participatory water resource governance in South Africa" *Geographical Journal* 177: 171–185.

Brown J. (2013) "Can participation change the geography of water? Lessons from South Africa" *Annals of the Association of American Geographers* 103: 271–279.

Castree N. (2003) "Commodifying what nature?" *Progress in Human Geography* 27: 273–297.

Connor J.D. and Kaczan D. (2013) "Principles for economically efficient and environmentally sustainable water markets: the Australian experience" in Schwabe K., Albiac-Murillo J., Connor J. and Hassan R. (eds) *Drought in arid and semi-arid regions* Springer, Netherlands, 357–374.

Dellapenna J. (2012) "The myth of markets for water" in Maestu J. (ed.) *Water trading and global water scarcity: international experiences* Routledge, Abingdon, 199–213.

Dryzek J. (2000) *Deliberative democracy and beyond: liberals, critics, contestations* Oxford University Press, New York.

EU (2012) *A blueprint to safeguard Europe's water resources* Communication from the Commission to the European Parliament, the Council, the European Economic and Social Committee and the Committee of the Regions, COM(2012) 673 final European Commission, Brussels https://www.eea.europa.eu/policy-documents/a-blueprint-to-safeguard-europes (Accessed 9 February 2018).

Evans M. and Pratchett L. (2013) "The localism gap – the CLEAR failings of official consultation in the Murray Darling Basin" *Policy Studies* 34: 541–558.

FAO Aquastat (2017) www.fao.org/nr/water/aquastat/tables/WorldData-Withdrawal_eng.pdf (Accessed 11 July 2017).

Ferguson J. (1994) *The anti-politics machine: development, de-politicisation and bureaucratic power in Lesotho* University of Minnesota Press, Minneapolis.
Goff M. and Crow B. (2014) "What is water equity? The unfortunate consequences of a global focus on 'drinking water'" *Water International* 39: 159–171.
Habermas J. (1962. 1989) *The structural transformation of the public sphere: an inquiry into a category of bourgeois society* Polity Press, Cambridge.
Hooijer A., Klijn F., Pedroli G.B.M. and van Os A.G. (2004) "Towards sustainable flood risk management in the Rhine and Meuse river basins: synopsis of the findings of IRMA-SPONGE" *River Research and Applications* 20: 343–357.
Lenton R. and Muller M. (2009) *Integrated water resources management in practice* Earthscan, London.
Li T.M. (2007) *The will to improve. Governmentality, development and the practice of politics* Duke University Press, Durham, NC.
Luks F. (2010) "Deconstructing economic interpretations of sustainable development: limits, scarcity and abundance" in Mehta L. (ed.) *The limits to scarcity. Contesting the politics of allocation* Earthscan, London, 93–108.
Mehta L. (2010) "Introduction" in Mehta L. (ed.) *The limits to scarcity. Contesting the politics of allocation* Earthscan, London, 1–8.
Mollinga P. (2010) "The material conditions of a polarised discourse. Clamours and silences in critical analysis of agricultural water use in India" *Journal of Agrarian Change* 10: 414–436.
Mollinga P.P., Meinzen-Dick R.S. and Merrey D.J. (2007) "Politics, plurality and problemsheds: a strategic approach for reform of agricultural water resources management" *Development Policy Review* 25: 699–719.
Monteiro P. and Matthews S. (2003) "Catchment2Coast: making the link between coastal resource variability and river inputs" *South African Journal of Science* 99, July/August https://researchspace.csir.co.za/dspace/handle/10204/766 (Accessed 9 February 2018).
Morley D. (2002) *Perspectives on freshwater. Issues and recommendations of NGOs* United Nations Environment and Development Forum, London https://www.ecolex.org/details/literature/perspectives-on-freshwater-issues-and-recommendations-of-ngo-s-mon-066127/ (Accessed 9 February 2018).
Muller, M. (2015) "The 'nexus' as a step back to a more coherent water resource management paradigm" *Water Alternatives* 8: 675–694.
Nickson, A. and Vargas C. (2002) "The limitations of water regulation: the failure of the Cochabamba concession in Bolivia" *Bulletin of Latin American Research* 21: 99–120.
NID (2016) *National inventory of dams* US Army Corps of Engineers http://nid.usace.army.mil/cm_apex/f?p=838:5:0::NO (Accessed 2 October 2017).
O'Neill J., Holland A. and Light A. (2008) *Environmental values* Routledge, Abingdon.
Reisner C. (1993) "Cadillac desert. The American West and its disappearing water" Pimlico, London.
Scientific American (2013) http://blogs.scientificamerican.com/plugged-in/2013/07/30/summertime-and-the-dams-come-down (Accessed 9 February 2018).
Shah T. (2010) *Taming the anarchy: groundwater governance in South Asia* Routledge, Abingdon.
Srinivasan V., Lambin E.F., Gorelick S.M., Thompson B.H. and Rozelle S. (2012) "The nature and causes of the global water crisis: syndromes from a meta-analysis of coupled human–water studies" *Water Resources Research* 48(10): 1–16.

Swyngedouw E. (2004) *Social power and the urbanisation of water – flows of power* Oxford University Press, Oxford.

Swyngedouw E. (2006) *Power, water and money: exploring the nexus. Human Development Report* Occasional paper, United Nations Development Programme, New York http://hdr.undp.org/en/content/power-water-and-money-exploring-nexus (Accessed 9 February 2018).

The Guardian (2006) www.theguardian.com/environment/2006/nov/01/society.travelsenvironmentalimpact (Accessed 9 February 2018).

Turner K., Morse-Jones S. and Fisher B. (2010) "Ecosystem valuation. A sequential decision support system and quality assessment issues" *Annals of the New York Academy of Science* 1185: 79–101.

Walker G. (2014a) *An uncooperative community: revisiting water privatisation and commoditisation in England and Wales* DPhil., University of Oxford.

Walker, G. (2014b) "Water scarcity in England and Wales as a failure of (meta)governance" *Water Alternatives* 7(2): 388–413.

WBCSD (2013) *Business guide to water valuation. An introduction to concepts and techniques* World Business Council for Sustainable Development, Geneva www.wbcsd.org/Clusters/Water/Resources/Business-Guide-to-Water-Valuation-an-introduction-to-concepts-and-techniques (Accessed 9 February 2018).

WEF (2014) *Water security: towards a values-based approach* World Economic Forum Global Agenda Council on Water Security, World Economic Forum, Geneva www.academia.edu/7819930/Water_Security_Towards_a_Values_based_Approach (Accessed 9 February 2018).

Woodhouse P. (1995) "Water rights and rural restructuring in South Africa: a case study from Eastern Transvaal" *Water Resources Development* 11: 527–545.

Woodhouse P. and Muller M. (2017) "Water governance – an historical perspective on current debates" *World Development* 92: 225–241.

WWC (2000) *World water vision: making water everybody's business.* World Water Commission Report World Water Commission/Earthscan, Marseille/London https://www.researchgate.net/publication/230557497_World_Water_Vision_Making_Water_Everybody's_Business (Accessed 9 February 2018).

WWC (2006) Summary of the Fourth World Water Forum www.worldwatercouncil.org/fileadmin/world_water_council/documents_old/World_Water_Forum/WWF4/synthesis_sept06.pdf (Accessed 9 February 2018).

WWC (2010) World Water Council www.worldwatercouncil.org/index.php?id=25 (Accessed 9 February 2018).

Young G., Dooge J. and Rodda J. (1994) *Global water resource issues* Cambridge University Press, Cambridge.

11 'Some are more equal than others'

Narratives of scarcity and the outcome of South Africa's water reform

Rebecca Peters and Philip Woodhouse

Introduction

Following the establishment of democracy in 1994, the South African national government embarked on a series of legislative reforms to provide a foundation for equitable social and economic development. While redress of historical injustice in access to land provided a mobilising force for the overturn of the apartheid government in South Africa, inequality in access to water was even more pronounced (Van Koppen 2008). The framework for water reform entailed two pieces of legislation: the 1997 Water Services Act (WSA) which is primarily concerned with provision of domestic water supply and services, and the 1998 National Water Act (NWA) which deals with broader water resources management. While there are important linkages between the acts, this chapter primarily focuses on water redistribution and reallocation associated with the implementation of the NWA. More specifically, it is about the values and social relations that influence the distribution of water, and their manifestation through narratives and metrics used to compare different uses and users of water.

Water values reflect the multiplicity of "cultural, social, symbolic, discursive, imaginative and ideational meanings, values and practices coupled with [water], which exist in complex ways alongside, and intermingle with, its basic material qualities as something that is essential for life, for agriculture and for society as a 'natural resource'" (Hart 2010, 34). Policies and practices related to water management, distribution, and access are never value neutral (Movik 2014) and the realisation of certain values for water over others is an expression of political and social power (Kemerink et al. 2011; Zeitoun 2014). Therefore, 'value' offers an approach that enables greater insight into the power relations embedded in water distribution than alternative avenues that view water policy and practice from an apolitical techno-managerial standpoint (Lankford et al. 2014; Meissner et al. 2013). Yet, much research regarding water governance and management tends to concentrate on either the physical or the managerial side of water, often avoiding questions of how political-economic power relations influence the physical and the managerial together in particularly, and invariably socially, uneven ways (Swyngedouw 2006). In this research we seek to critically examine how and which values are expressed in policy and practice, rather than solely discussing the procedural aspects of water governance and management (Movik 2014). This approach to the study of value facilitates a close examination of the competing claims of value and the associated 'aggregate evaluation' of their consequences. In this respect, the 1998 NWA in South Africa is a useful case for analysis of the expression of values, since, unlike most water legislation, it expressly set out to redress historic patterns of inequitable distribution of access to water arising from racial discrimination in South Africa.

The central question this research seeks to answer is: how do values of water expressed in the 1998 NWA overlap and differ across institutions, users, and intended beneficiaries in practice? Three interrelated lines of inquiry are explored. The first examines the kinds of discourses employed by different institutions and actors. The second considers the role of different metrics used to compare the value of water consumed by different types of water use such as volume available for irrigation, flow levels in a river, or regularity of supply. Examining the measurements used to characterise water for domestic use or for irrigated agriculture illustrates how certain 'techniques of value' overlap and also when they differ. The third line of inquiry concerns the institutional assemblages through which value is articulated both formally and informally between and among different institutions and users. Collectively, these lines of inquiry examine how value is constituted, why institutions align behind different measurement criteria and valuation approaches, as 'benchmarks' for competing claims. Drawing from this analysis, the concluding section discusses lessons that may be drawn from the experience of water reform in South Africa.

We draw on a study of water allocation reform in Mpumalanga province, and the Inkomati Usuthu Water Management Area (IUWMA), one of nine water management areas established by the Department of Water Affairs (DWA, formerly DWAF) to implement a decentralised model of water resource management as articulated by the framework of the National Water Resource Strategy (DWAF 2004). In the IUWMA irrigated agriculture on white-owned farms is the historically dominant water user, accounting for approximately 57 per cent of all water requirements in the catchment (Woodhouse 2012, 855). Other major users of water are pulp mills, forestry, and environmental interests such as the Kruger National Park. Domestic water use accounts for around 10 per cent of the total. The Gini coefficient (0.95) indicates greater inequality for direct water use than for income (0.64) between white and black populations (Van Koppen 2008). This reflects the legacy of the "historical exclusion of the black majority population from property rights in water under the apartheid regime" (Woodhouse 2012, 848). To address these disparities, the 1998 NWA mandated the creation of two user-driven water resource management organisations (Faysse 2004, 52). More inclusive WUAs at the local level would replace the irrigation boards (IBs) constituted during the apartheid years by white commercial farmers.

Catchment Management Agencies (CMAs) at the larger catchment scale would serve to decentralise water governance and decision-making and create channels for representation of the main water-using sectors. Within the IUWMA, the Inkomati Catchment Management Agency (ICMA, more recently IUCMA with the inclusion of the Usuthu area), was the first CMA to be established in 2004 – in recognition of its status as a 'stressed' catchment.

The analysis draws on a number of previous studies (Woodhouse 1995; Woodhouse and Hassan 1999; Brown 2011, 2013; Woodhouse 2012) updated by further data collected by Peters (2015) through semi-structured interviews with CMA members, IB members, environmental scientists, a South African National Parks (SANParks) manager, municipal water managers, academic researchers from the Institute for Poverty, Land, and Agrarian Studies (PLAAS), in the University of the Western Cape, commercial farmers, and 'emerging' farmers in former Bantustans.

Water reform and value in South Africa

The 1998 NWA: foundation for reform

The 1998 NWA was strongly guided by new post-apartheid national values "such as equity, redress of past wrongs, poverty reduction, environmental sustainability and economic and financial sustainability" (Merrey 2011, 252). The 11 goals of the NWA are as follows:

- a meeting the basic human needs of present and future generations;
- b promoting equitable access to water;
- c redressing the results of past racial and gender discrimination;
- d promoting the efficient, sustainable and beneficial use of water in the public interest;
- e facilitating social and economic development;
- f providing for growing demand for water use;
- g protecting aquatic and associated ecosystems and their biological diversity;
- h reducing and preventing pollution and degradation of water resources;
- i meeting international obligations;
- j promoting dam safety;
- k managing floods and droughts.

Schreiner (2013) posits that the NWA is premised on "balancing the three legs of social benefit, economic efficiency and environmental sustainability" while outlining "the legal framework for the national government to protect, use, develop, conserve, manage and control water resources" (Schreiner 2013, 240). In essence, Schreiner identifies equitable access ('social benefit'), development ('economic efficiency'), and sustainability ('environmental sustainability'), as well as security ('legal framework'). However, while these values are expressed in the NWA, the explicit, clearly articulated values need to be disentangled from tacit value assumptions to enable a coherent understanding of their influence on water governance and management policies and practices. Hereafter we focus on the normative categories of equitable access and development to understand how some values become dominant and gain power while others are suppressed or ignored.

Equitable access

The value of 'equitable access' in the NWA is most closely associated with the first three of its goals, particularly "promoting equitable access to water" and "redressing the results of past racial and gender discrimination". Redress of historical injustice is closely associated with water reallocation and redistribution mechanisms. The mandate to create two user-driven water resource management organisations suggests an improvement in equality of decision-making (Faysse 2004, 52): WUAs to replace the historically racially exclusive IBs and CMAs at the larger catchment scale.

Economic development

Economic 'development' values may be identified through the next three goals of the NWA, particularly "facilitating social and economic development". Economic development is primarily concerned with how water resources management may help 'the economy' or the economic benefits of particular irrigation scheme projects. Economic considerations predominately relate to which values and costs 'count', yet do not create clear intra-category means of comparison for different kinds of economic values. Notably, although the NWA repealed apartheid-era laws and abolished private property rights in water, a category of 'existing lawful use' (ELU) allowed commercial farmers to continue prior levels of use of water resources. The 1998 NWA sought to accommodate a diversity of new water users and values across the transitioning country.

An important aspect of legitimacy of the NWA was that it incorporated principles espoused at the 1992 International Conference on Water and the Environment (ICWE) in Dublin. The three 'Dublin principles' of water governance evident in the NWA – decentralisation, participation, and economic pricing – were, at the time, widely promoted as international benchmarks of best practice in water governance. The NWA prioritises *decentralisation* through "the principle of subsidiarity – management of water resources at the lowest appropriate level, through catchment management agencies" (Schreiner 2013, 240). Decentralisation is associated more generally with the idea of providing bounded hydrological units (catchments) within which use of 'finite' water resources should be managed. It was also widely held that decentralisation has the "potential for democratic *participation*, in which local actors could mount challenges to national and local elites"

(Postero 2007, 126). In South Africa, the NWA therefore provided a formalised set of processes for rural black communities to engage in decision-making and challenge the status quo of prioritisation of water allocation. The principle of economic value as expressed by pricing for water in the Dublin principles is reflected in Section 5 of the NWA that all water use is subject to a water resources management charge. The stated purpose of these charges is to fund the operation of the CMA. The Dublin principles are underpinned by a more fundamental narrative of water as a 'scarce' resource: finite and subject to competing demands. This resonates with the official designation of the IUWMA as a 'stressed' catchment in which water use needed to be contained or reduced due to high levels of existing water usage relative to flows in the rivers, particularly in the dry season.

Water users in the IUWMA

Both before and since the NWA, irrigated agriculture has been the dominant water user in the IUWMA, accounting for approximately 57 per cent of all water requirements in the catchment and up to 73 per cent of water demand in individual sub-catchments (Woodhouse 2012, 231). The "geography of irrigated agriculture still largely reflects its development under the apartheid era" and the vast majority of commercial farmers are white (Woodhouse 2012, 232; Bernstein 2015). However, there is also significant demand for irrigated agriculture among black farmers, including small-scale irrigators producing vegetables, small- to medium-scale sugar-cane 'out-growers' under contract to sugar mills, and the 'beneficiaries' of land restitution claims who have become owners of irrigated plantations under South Africa's land reform programme (James and Woodhouse 2017).

Large-scale commercial forestry plantations dominate the higher altitude (and higher rainfall) parts of the Mpumalanga landscape. Accounting for an estimated 11 per cent of total water requirements in the IUWMA through interception of rainfall and stream flow reduction, forestry is barred from further expansion by the DWAF (subsequently the DWA) (Woodhouse 2012, 233). Mining and industry are not differentiated by the DWA or IUCMA, yet are a significant user of water. The primary users are ESKOM and a Sappi Kraft paper mill in the upper catchment, and two Transvaal Sugar Limited (TSB)[1] mills in the lower catchment (Bezuidenhout and Singels 2007).

'Domestic' water use for drinking, washing, and other household demands is not homogenous. It varies across rural and urban contexts, as well as with the socio-economic arrangement of the household. This chapter will accordingly consider the diversity of domestic water use, including for subsistence farming and small-scale irrigated agriculture. Such 'primary' use of water is governed by separate legislation under the 1997 WSA, rather than the 1998 NWA.

Maintaining river flows for ecological goals was one of the innovative features of the NWA. The 'environmental reserve' or water to maintain ecological flow is significant within the IUWMA, largely due to the Kruger National Park and Sabie-Sand Game Reserve (Woodhouse 2012, 234). The Kruger and Sabie-Sand are downstream of the main irrigation and forestry users on the Sabie-Sand and Crocodile Rivers. Therefore, if flow levels entering Kruger are too low due to overdraw from other users or drought, the "ecological functioning of the park will be affected" (Woodhouse 2012, 234). A further downstream constraint on water use is the commitment to international agreements to share water (12 per cent of the Crocodile sub-catchment flow and 20 per cent of the Komati sub-catchment flow) by maintaining cross-border flows into Mozambique (Zeitoun et al. 2014).

NWA implementation: institutional framework

We need to briefly outline the institutional participants in implementing the NWA, bearing in mind disparities between formal legislative objectives and capacity, operation, and function in practice that will be analysed in subsequent sections.

IUCMA

Within the guidelines of the National Water Resources Strategy, each water management area (WMA) is to be managed by a single CMA. Within the Inkomati Water Management Area, the IUCMA was the first to be established in 2004. However, the IUCMA struggles to implement coordinated and cooperative participatory management to ensure water resource use is sustainable, equitable, and efficient, as required by the NWA (Tewari and Oumar 2013).

DWA – formerly DWAF

As CMAs were tasked with undertaking delegated water management activities within given WMAs, DWA was assigned responsibilities for four sub-directorates including water quality and pollution control; water resources including dams, infrastructure, and allocation to IBs; and planning and development tasks including new projects associated with national development priorities (DWAF 2005a). Unlike the CMA, the DWA's purview is only within political rather than 'watershed' boundaries for responsibility. As will be discussed, the perception by DWA of losing power to the IUCMA has generated significant tensions between the institutions.

IBs and WUAs

IBs, previously established under the 1956 Water Act, were responsible for water distribution for white farming interests. The NWA obliged IBs to transform into "voluntary" WUAs, setting quotas for blacks and women to comprise 50 per cent of each management committee (Schreiner and Hassan 2011, 29). Operating at a strictly local level, the primary purpose of WUAs is not water management but rather to act as a cooperative association for individual water users wanting "to undertake water-related activities for mutual benefit" (Woodhouse and Hassan 1999, 32; Hassan and Farolfi 2005). WUAs are supposed to be formed from existing IBs as prescribed in the NWA, which sets out the procedures for establishment and operation of these associations (Woodhouse and Hassan 1999, 32). However, a number of issues to be described below have prevented the transformation of IBs to WUAs.

Municipalities

Within the province of Mpumalanga, there are three district municipalities further divided into 18 local municipalities. Under the 1997 WSA, legislative duties were assigned to municipalities to provide water supply and sanitation according to "national standards and norms" (Muller 2008, 146). Therefore, municipalities are supposed to serve as 'water service authorities' to fulfil a regulatory mandate with respect to water service providers. However, some municipalities also serve as service providers, leading to potential conflicts of interest (Muller 2008). Municipalities are responsible for purchasing bulk water from DWA to supply to their residents (Nieuwoudt et al. 2004).

Traditional authorities

Traditional leaders maintain control in many former homeland areas in the IUWMA (Malzbender et al. 2005; Claassens 2015). The separate institutional reform process in former Bantustans via the Department of Agriculture (DoA) means that these areas are "isolated from irrigation boards (future WUAs)" as well as the statutory process of registering and licensing water use (Woodhouse 2012, 865; Malzbender et al. 2005; King 2007).

Discourse in the NWA

Within the institutional landscape set out above, implementation of the NWA has been interpreted through different narratives of the goals of water allocation and its reform. Here we outline three distinct narratives encountered among water users and the institutions through which water use is governed: scarcity, participation, and rights.

Scarcity discourse: efficiency, productivity, and pricing

A discourse of scarcity links easily with the official designation of the IUWMA as 'stressed' but also underpins metrics of efficiency and productivity that in practice constitute 'ability to pay' as a criterion for water allocation.

The representation of certain users as 'efficient' or 'wasteful' of scarce water has played an important role in maintaining established distribution of water use. In practice three elements combined to produce this result.

First, a concept of 'existing lawful users' (ELUs) was coined by a lawyer advising the drafters of the NWA in order to forestall litigation by white farmers and the possibility of economic disruption if existing water rights were simply cancelled (Movik 2014, 191). The caution surrounding ELU is evident in the guidelines in the 2005 draft water allocation reform document:

- ELUs will only be curtailed as a last resort and only after all other options to find water for the poor and [Black Economic Empowerment] have been exhausted;
- ELUs of water will not be curtailed unless there are clear procedures and support programmes established to promote the productive use of water by emerging [black] farmers;
- No existing lawful consumptive user of water will be completely curtailed.

These guidelines suggest a presumption that ELU "are making productive, efficient and beneficial use and are contributing to socio-economic stability and growth" (Movik 2014, 191).

In the IUWMA a second scarcity-linked discourse frequently encountered is "all of the water has been allocated". Taken together, these serve not only to legitimate ELU (and particularly white commercial farmers) as productive and legal but also to foreclose further discussion of new water allocation within 'existing' resources administered within the catchment. This confirms the concerns expressed by some policy-makers and scholars (Schreiner and Van Koppen 2002) throughout the reform process that defining particular users as 'lawful' would prevent meaningful redistribution (interview with CMA legal advocate, June 2015). The NWA explicitly emphasised demand management (i.e. restricting overall use of water), through tariffs and water charges, as well as local water rationing, as strategies to compel CMAs to reconcile the water resource of a catchment with the demands of development. Yet the sense that all the resource has already been allocated to existing lawful (and productive) users both pre-empts arguments for reallocation and places any 'supply-side' interventions (e.g. by building more dams to store rainy season floodwater) to increase the supply of water to 'historically disadvantaged individuals' (HDIs) and 'emerging users' (black farmers) as being outside the remit of the IUMA. This places the burden on underfunded municipalities without a tax base to finance new infrastructure to supply their own water. An interview with a CMA administrator revealed an implicit underlying attitude: "if urban black areas want more water, they have to increase their own supply" (June 2015).

This separation of the needs of redress and equitable access from maintenance of ELU is reinforced by a third discourse that constructs HDIs as wasteful, disruptive, and potentially ill-equipped to manage scant water resources responsibly (Movik 2014). This is evident even in official documents: "if [water] reallocations occur too quickly, the country will suffer economic and environmental damage as emerging users struggle to establish productive uses of the reallocated water" (DWAF 2005b, 4).

Participation discourse

Participatory processes were seen in the ICWE principles as an important counter to economic valuation, and ensuring locally relevant decentralisation of "democratic decision-making procedures as an element and condition for social justice" (Young 1990, 23). The three assumptions underpinning participation discourses regarding the role of the state, the universality of application, and the transformative potential of institutional reform as discussed by Brown (2011) are thus usefully supplemented with a consideration of decentralisation. Critically, while the NWA views CMAs and WUAs as 'more inclusive' through ensuring a wider range of 'stakeholders' in water use, commercial agriculture's IBs viewed WUAs as threatening established commercial farming interests (interview with IB Secretary, June 2015). Thus, while the NWA changed the formal rules of participation through the establishment of CMAs and WUAs, the "tenacious survival capacity" of institutions with strong economic interests such as IBs have resisted change (Brown 2011). In practice, therefore, a discourse of participation is accepted only insofar as it generates decisions conforming to the logic of scarcity and the need to protect 'ELUs'.

Rights and citizenship discourse

A very different 'rights-based' discourse emerging in rapidly urbanising former Bantustans and other deprived areas in the IUWMA draws upon earlier narratives regarding non-payment (King and McCusker 2007). There is a range of views in the current debate within the water policy community regarding the origins of non-payment, and these debates are useful to understand the context of interviewee responses. One line of debate suggests that the 'culture of non-payment' stems from the anti-apartheid rent boycotts in the 1980s (von Schnitzler 2008). Following the first democratic election in 1994, the ANC prioritised the resumption of payment for services. The renewed imperative of paying for services was embedded in language of 'empowerment' and 'active' citizenship, as exemplified by a 2004 *White Paper on Water Supply and Sanitation Policy* (DWAF 2004; von Schnitzler 2008).This stated a need to confront the 'sense of entitlement' of disadvantaged people and their negative 'attitude toward Government policy' regarding water pricing. However, it contradicts a 'perception' that it is the 'Government's responsibility to provide water services' as it is stated in the Constitution, and a series of government campaigns in former Bantustans seeking 'nation building' through 'responsible citizenship' have failed to reduce rates of non-payment (von Schnitzler 2008, 907; interview with emerging farmer in former Bantustan, June 2015). Another argument suggests that people believe the (free) provision of water is a result of liberation and democratic rule (Muller 2007). The South African Constitution contains the "right to have access to… sufficient food and water" (Section 27b) and requires that the state takes "reasonable legislative and other measures within its available resources, to achieve the progressive realisation of each of these rights" (SA Constitution). In practice, the government has conceded a right to 25 litres per person per day "Free Basic Water" – a limit that is heavily contested by the poor (von Schnitzler 2008).

Underlying the discourse of rights to water is the historical record of centralised state planning and heavily state-subsidised water infrastructure for white towns and white commercial farmers under apartheid rule (Van Koppen 2008; Bernstein 2015). The tendency of water policies and institutions to obscure the history of past subsidisation while demanding that poor users pay for water, and for any investment to increase their access to it, contributes to the current discourse on the part of municipalities and black residents that (free) water should be a "gain of liberation, not a commodity" (interview with resident in former Bantustan, June 2015). Conversely, however, the benefits of past subsidisation of urban and rural infrastructure in wealthy white areas is manifested not only in accumulated wealth but also in capacity for water management, as compared with black areas which lack that capacity and whose people are characterised as inefficient water users.

Calculability: metrics and the performance of valuation

Discourses of scarcity are materialised ('performed') through calculations of water quantity and quality used every day by the CMA and DWA regarding water value within the IUWMA. These concern decisions about water licensing determinations, assessing ELU, tariff structures, infrastructure construction, and establishing the flow of water needed to sustain the ecological health of a river. Calculations and calculability are therefore intimately related to water value. This section will explore why institutions prioritise certain methods of calculability of water, how this increases power or authority of some values in water relative to others, and the ways that calculations of water values are contested in practice.

The IUCMA does not administer water for individual domestic users as it only measures bulk supply to municipal governments responsible for purchasing water. Driven by population growth across both rural and urban areas, demand for expanded and improved water supply means that municipal politicians pressure the DWA and CMA to increase water supply to their districts for domestic consumption. This is resisted by the CMA whose advice and technical guidance is to wait until supply can be improved (interview with CMA engineer, June 2015), which invariably means new (and large) storage reservoirs to be financed by central government. In resisting municipalities' demands an important 'calculative device' is, in fact, the non-measurement of water used, which allows assertions of over-use as needed. Thus, although all water uses are supposed to be metered throughout the catchment, many are not, and "almost all municipalities use much more than their given allocation" (interview with CMA engineer, June 2015). Thus, in the absence of improved piped drinking water access and an increase in illegal connections for domestic use, primarily in 'informal settlement' and former Bantustan areas (interview with IWR Water Resources engineer, June 2015), the CMA estimates economic losses to municipalities of around 70 per cent from non-payment and illegal connections (interview with CMA engineer, June 2015). A consequence of municipal "inability to deal with the illegal connections" is then that these areas will not be "given an increased allocation" because "municipal leadership must first invest in improving systems and infrastructure" (interview with CMA official, formerly with DWAF, June 2015).

The 'problematic' character of supplies of water to growing populations, the majority of which are black, is further underlined by the way 'domestic' (or 'primary') water use is referred to as 'high impact'. Conversely, commercial irrigated agriculture is viewed by the IUCMA as being more flexible in seasonal consumption than domestic users, and willing to put limits on use in dry times as the result of functional community driven IBs or WUAs (interview with IB secretary, June 2015). This use of 'assurance of supply' (typically 80 per cent for commercial agriculture, 100 per cent for domestic use) is used to justify a 'lower' impact attributed to irrigated agriculture despite it being responsible for over 55 per cent of total volume of water used, compared to around 12 per cent for all domestic use in the catchment. A similar metric is evident in contrasts drawn between large- and small-scale farmers: "most commercial farmers save water for the next year and view this as a good thing, big businesses want long-term continuity of operation and are willing to cut use in one year to have assured supply for the next. But black farmers are more likely to take risk, over-using an allocation and get cut off [from supply] than commercial farmers" (interview with IB Secretary, June 2015). This perception of commercial farmers saving water and emerging farmers taking risks suggests a notion of 'responsibility' regarding calculability – the long-term responsibility of an established farming organisation, although this obscures the pressures and uncertainty on black farmers.

In practice, the lack of systematic measurement of water use is not only used as a cover for assertions of wasteful use by black populations, but also as a means to obscure potential illegal overuse of water by commercial farms. Thus the use of metrics for water use is critical to the ability of the commercial farming sector to exercise de facto control of water. In determining pricing and allocation, the volume of water abstracted or withdrawn from a river is the metric used by the IUCMA. Under the previous 1956 Water Act, water was allocated according to land area, but the NWA introduced volumetric measurements. However, the capacity of municipalities and the IUCMA to calculate withdrawals varies across the catchment area. For example, while some users taking water from the Crocodile River area use meters, others – without sufficient funding or technical capacity to meter – do not (interview with CMA engineer, June 2015). Therefore, there is "no precise measure of abstraction" and "less than 40 per cent of all water use in the catchment is measured" according to a CMA river operations manager (interview with CMA engineer, June 2015). This meant that DWA efforts to conduct a verification of water use as a prelude to water allocation reform was based on satellite images to estimate areas irrigated.

However, using the areas irrigated to determine the volume of water abstracted from rivers depends on estimates of efficiency of water use, and this proved a barrier to the government's ability to verify water use by farmers. In practice, the DWA simply issued licences based on the amounts of water commercial farmers claimed they were using (Woodhouse 2012). Current practice relies on a model designed by an engineer with a local private consulting firm using data from the Food and Agriculture Organisation for individual crop-type requirements for water, assuming optimum water use, to provide a baseline. Actual water use is then estimated by taking account of variable factors such as rainfall and availability of stored water to determine operating rules (interview with IWR Water Resources engineer, June 2015). The current model used for both irrigated agriculture and forestry, called SAPWHITE, has become increasingly refined in its methodology. However, gaining access to inputs for this model and calculating water needs for different crops depends on a degree of technical capability, and access to institutions (interview with municipal manager, July 2015).

This acts as a significant barrier to 'emerging' farmers in seeking irrigation in former Bantustans (Cousins 2011). Most significantly, encounters between those who are existing large-scale water users and those who are seeking to increase their water use are hugely asymmetric in terms of knowledge of how much water ('legal' and 'illegal') is being used and by whom. Variation in transparency and knowledge resulting in differences in access to calculability between users contributes to the emergence of "ambiguously defined responsibilities and an imprecisely defined accountability" (Cousins 2011, 62) within water governance and management structures in the IUWMA, leaving many poorer farmers without reliable access to water or ability to participate in decision-making about resource allocation. While some of the information asymmetry is inherited from the pre-NWA period (Woodhouse and Hassan 1999), little improvement for irrigation pump operation has resulted, despite availability of digital equipment (Woodhouse 2012, 858).

Yet the IUCMA's conceptualisation of 'political interference' by municipalities maintains water firmly in a domain where the political values of technocrats become the norm and 'interference' is viewed by the IUCMA as aggravating the formalised processes in which it approaches participation. Nonetheless, individuals living in areas lacking irrigation and drinking water systems voice their dissatisfaction with the lack of change, including with their political representatives: "politicians want to put more pump stations in to further short-term political careers, yet people do not get water" (interview with emerging farmer, July 2015). Thus far, the extent to which total available water is used by 'ELUs' sets the terms ('all water is already allocated') by which water available for reallocation from commercial farmers (ELU) to emerging farmers (HDIs) can be discussed. Inasmuch as the success of water reforms outlined by the NWA are premised on verification of 'ELU' as a baseline for reallocation and licensing, water reallocation was effectively stalled (Woodhouse 2012, 858).

The inability to use technical metrics to reallocate water is evident even in instances where considerable scientific expertise is deployed, as in the case of calculating the minimum flow, or 'reserve', that is needed to maintain ecological functioning of rivers, a requirement also introduced by the NWA. While scientists and water managers developed ground-breaking determinations for ecological flows, DWAF staff found it difficult to translate monthly flow requirements into a format required to issue 'licences' (interview with environmental scientist consultant, June 2015). Furthermore, monitoring plans established by the CMA were insufficient to determine if ecological flows are maintained or not (interview with CMA engineer, June 2015). Dams in the IUWMA are not designed to allow controlled releases, rather spilling or having a constant outflow (interview with IB Secretary, June 2015). Thus, despite the innovative determinations from scientists for flow requirements, none of the rivers in the IUWMA catchment met stated reserve levels (Pollard and du Toit 2011).

The completion of verification and licensing of water use may be a basis, in principle, for reallocation. However, based on ELU, the DWA has repeatedly stated "no new water licences will be possible" (interview with DWA municipal manager, July 2014). Particularly since no quantitative targets were set for water reallocation, the refusal to issue new licences based on discourses of scarcity, perceptions of emerging farmers as 'harmful' or 'less beneficial users' as compared with commercial farmers has allowed the calculative power of commercial agriculture to maintain authority (interview with PLAAS researcher, June 2015) at the expense of 'equitable access'. The authority of commercial agriculture and its IBs may decline over time as price structures are expected to shift in the coming years and this is further increasing tension between the DWA, IUCMA, and commercial farmers (interview with IB Secretary, June 2015). Thus, commercial farmers are anticipated to have an increasing burden to show economic and employment viability, as more red flags are raised regarding the high percentage of water they use and low price they pay for it (interview with DWAF consultant, June 2015).

Institutional assemblages as barriers to change

We have so far focused on the values that are manifest in discourses of water allocation and the calculative devices through which they are 'performed'.

We now turn to the institutional frameworks and alliances within which these processes have been played out. We identify four distinct, though intersecting, institutional dynamics: the withdrawal of state agencies in favour of those of markets; the efforts to decentralise water management; the urbanisation of former Bantustan 'homeland' areas; and the processes of land reform.

State withdrawal

Under the NWA, the state acts as the custodian of water in the public interest and the national political economy is critical for analysis in relation to water governance and management. Departing from the pro-poor strategy of the basic needs-oriented Reconstruction and Development Programme (RDP), the ANC adopted the 'market friendly' Growth, Employment, and Redistribution (GEAR) programme in 1996 (Peet 2002). Whereas the RDP contained concrete targets for redistribution, goals for poverty alleviation, and explicit statements regarding the role of government in adopting pro-poor policies (Weeks 1999), GEAR focused on export-oriented industrial expansion, privatisation of state utilities, and trade liberalisation (Peet 2002). This programme prioritising strong growth as the basis for poverty and unemployment reduction (Weeks 1999; Gelb 2007) did not "mention reducing inequality as a policy goal" and removed proposals to increase re-distributional taxes (Weeks 1999, 6). Consequently, GEAR committed the ANC to a long-term path of economic policies prioritising growth over more equitable redistribution (Cousins and Walker 2015). It is within this political economy context that the initial ambitions of the NWA regarding water reform came to prioritise productivity and economic efficiency rather than poverty reduction and rural livelihoods (Hall 2015, 135; Tapela 2008). More concretely, the policy was associated with a dismantling of state subsidies to agriculture and a rapid consolidation of the agricultural industry into larger corporate commercial units (Bernstein 2013).

The shift in emphasis from state to corporate commercial agencies during the 1990s saw a transfer of expertise from DWAF to private sector engineering companies, many of which were also subject to consolidation and takeovers by foreign multinational companies. Similarly, the extension and technical experts that had been employed within the Bantustan administrations were formally transferred to the newly formed Mpumalanga DoA, but in practice also ended up as individual consultants, or in a variety of non-government agencies that were subsequently contracted by government to provide technical services. A key player from the point of view of water services was the sugar company TSB (now RCL Sugar) which had contracts with some 1,600 black 'out-grower' sugar-cane producers in ex-homeland areas, and consequently took over much of the technical and financial advisory work previously provided by government (James and Woodhouse 2017).

In 1994, the outflow of technical experts from DWAF to private consulting, driven by racial tensions as well as clashing political perspectives, was concurrent with a push to hire black staff in DWAF. Using the analogy of 'fixing a plane while in flight', Schreiner (2013, 241) explains the negative cycles contributing to the occurrence of increased technical demand and decreased capacity of staff. From a techno-managerial perspective, the primary challenges faced by the DWAF in the following decade included "lack of experience, limited technical knowledge, and high staff turnover" which were "worsened by the transfer of skills to the private sector" from the DWAF (interview with former DWAF staffer now at the IUCMA, June 2015). This de facto privatisation of public sector expertise meant also the removal to the private sector of a network of professional experience only accessible to incoming staff in DWAF via commercial contracts. DWAF thus came to depend on, and pay for, private consultants to 'support' the implementation of the water reforms while the responsibility for 'actual implementation' remained with DWAF.

Decentralisation

The proposed decentralisation under the NWA of water resource management through creation of CMAs initially met strong engagement by commercial farmers and their IBs. A consultation process in the three sub-catchments of the then-IWMA (Sabie, Komati, and Crocodile) sponsored by the local DWAF officials resulted in a proposal for a CMA that was submitted to DWAF in Pretoria but rejected as being insufficiently representative of historically disadvantaged groups (Brown and Woodhouse 2006). This initiated a prolonged struggle in which the DWAF in Pretoria sought to impose a reform of water allocation through creation of the ICMA (subsequently IUCMA). The DWAF nominated a governing board of 14 members in 2005 but the CMA was only operationalised in 2010 with the launch of the Catchment Management Strategy. The interval saw attrition of members of the governing board, departures of key personnel, and a slow accretion of capacity, mostly through recruitment of staff from DWAF. Throughout, progress towards reform of water use has been blocked by poor relations between the CMA and the commercial farmers' IBs (Woodhouse 2012). As a consequence, elaborate technical efforts to verify the levels of existing water use by commercial farmers have failed to do more than convert 'ELU' into formal licences, despite evidence of much higher (and therefore illegal) water use by farmers. Similarly, the goal of converting IBs into 'water users' associations' that are more representative of all water users sharing a particular infrastructure or stretch of river remains largely unmet (Woodhouse 2012).

The barriers for transformation include resistance from IBs, on the grounds that they did not want to be responsible for collecting fees from (black) users who refused to pay them. IBs also claim that their proposals to convert to WUAs were either denied or never responded to by the DWAF (interview with IB Secretary, June 2015; Seshoka et al. 2004). The standoff with the commercial farmers is particularly problematic as the payments for catchment management charges, which are the basis of CMA budgets, have been retained by IBs, thus undermining efforts to maintain and improve water delivery infrastructure to include historically disadvantaged users. The reduced capacity of local DWA is alleged to have resulted in some licence applications sitting in the DWA for eight years without being finalised (interview with DWA manager, July 2015). The experience of the CMA raises questions about whether water governance models using permits and licensing, often viewed as "modern" and "progressive", are rather a bed of inefficiency (Tewari and Oumar 2013). Schreiner (2013) similarly suggests that the slow roll-out of licences is due to technical approaches to registration of water use that are both inadequate for billing and revenue collection, and also do not ensure compliance with registered use. This widely shared concern that the NWA is not geared to resolve distribution issues in a timely manner is furthered by the high administrative costs of implementation, poor incentives for long-term investments, bureaucratic inefficiency, and practical problems in water pricing for the poor. That the overall result of a decade of decentralisation is that no new licences can be allocated because pre-existing ('lawful') water use has been ratified by the new institutions suggests a more fundamental problem with the reforming potential of the decentralisation model. It may seem unsurprising that commercial farmers used the narrative of water scarcity and its concomitant discourse of prioritising 'efficient and productive' use to protect their domination of water allocation. Their capacity to do this, however, may have been strengthened by the creation of a 'sectoral' assemblage in which 'local' appropriators of water could mobilise against 'imposed' and 'disempowering' reforms by a central government agency. A more fruitful approach may have been to bring the decentralised governance of water under the local provincial government of Mpumalanga and to thus engage water allocation more directly with local democratic (albeit flawed) processes.

Urbanisation

This last proposition is underlined by rapid changes in the character of settlements in ex-homeland areas of Mpumalanga. The apartheid government vested formal authority for water resources management in homeland governments represented by appointed traditional chiefs and councils (Van Koppen et al. 2002). These forms of autocratic leadership come into tension with the NWA's recognition of the primacy of democratically elected local governments and public participation (Malzbender et al. 2005). While the 1956 Water Act and other related acts were applicable across territories of the nation-state and self-governing homeland territories, because Bantustan farming was categorised as subsistence farming, it did not fit with the dominant perception of economically productive and therefore was not prioritised. Formal ownership and management of both land and water within Bantustan irrigation arrangements also remained with the state, with traditional authorities playing a de facto role for community land reallocation (Van Koppen et al. 2002). Despite the introduction of elections and democracy in 1994 a dual political system is entrenched in many former homelands with tribal authorities. Exemplifying this, the 2013 Rural Development Framework of the Department of Rural Development and Land Reform (DRDLR) transferred communal land use rights to traditional councils, rather than to "ordinary rural people who will obtain only 'institutionalised use rights' subject to the decisions of non-elected traditional leaders" (Cousins and Walker 2015, 79).

However, areas under traditional authority are also caught up in processes of social and economic change accompanying urbanisation that may ultimately transform political structures (Claassens 2015). The common 'fault line' between urban and rural water needs and values (Hope and Garrod 2004; Claassens 2015) may not be useful in the context of rapidly changing patterns of land use and settlement. Many of these areas have since merged under unified local administrations (interview with AWARD researcher, July 2015). The density of newly urbanised areas bringing investment and business has altered the social and political economy of deprived areas into zones of consumption (interview with municipal authority, June 2015; Claassens 2015). The shifting urban expectations accompanying internal development contribute to new forms of contestation regarding the role of the state and municipality to provide water to marginalised communities. Further, since the NWA established municipal authorities as responsible for water service provision, some (rural) areas governed under traditional authority are now seeking to become formally part of a municipality (interview with AWARD researcher, July 2015; King and McCusker 2007) due to frustration with traditional authorities not working with municipal government to improve services. By becoming a 'formal' part of the municipality, the hope is that better access to services may result (interview with 'smallholder' farmer in former Bantustan area, June 2015).

Land reform

Some have stated a necessary linkage between land reform and water reform (Merrey 2011, 254), especially since ownership of rights in either resource makes it possible to acquire control over the other. Yet the legislation, policy, and organisation of land and water reform in South Africa have been quite distinct (Woodhouse 2012). An exception to this is evident in the settlement of land restitution claims by communities displaced from their land in the Mpumalanga Lowveld during the 1950s. In 2008 the claims were settled, transferring to black communities a total of 61,202ha for which the South African government paid the incumbent white farmers R2.8 billion ($426.7 million).[2] Much of the land was commercial irrigated sugar cane, citrus, or banana plantations. The owner of the sugar-cane mills, TSB, had encouraged landowners not to contest the land claim, ensuring that the claim was settled quickly. It also moved swiftly to persuade the DRDLA to put in place arrangements to farm the land through a number of 'joint venture' companies in which community trusts, to whom landownership had been transferred, were partnered with the company. As a consequence, the sugar-cane company now farms directly more land than it did before the land reform, although this arrangement appears to offer little benefit to the majority of African 'land-owners' (James and Woodhouse 2017).

From the point of view of water reallocation, a key consequence of this transfer of land rights of large areas of irrigated farmland, and their existing associated water rights, to the nominal ownership of black communities is widely asserted as obviating the urgency for any allocation of new water licences to 'emerging' black farmers. The need for government agencies to demonstrate efficient and productive use of land following land reform has here reinforced the narrative of efficient and productive use of water promoted by corporate agriculture seeking to retain access to water to maintain its supply of agricultural raw material to the sugar mills. In so doing, alternative, smaller-scale uses of both land and water by black people have been foreclosed. A key institutional element of this assemblage is the 'collective' status of both land and water ownership by black people which, contrary to the 'individual enterprise' ethos of much of South Africa's economic policy, perpetuates the property rights regime allocated to black people under apartheid. This subordinates individual use of land and water to the 'collective' embodied in traditional leaders. Aside from problems of unaccountable and rent-seeking behaviour of such authorities, such collective rights impose barriers for individuals seeking to develop economic activity which invariably involves engagement with South Africa's fiscal authorities and financial institutions, as well as agricultural markets.

More broadly, the deregulation of water and land allocation to favor 'economic development' values associated with commercial irrigated agriculture suggests a racialised interpretation of deregulation of productive activity. As Movik (2012) points out, the state's emphasis on creating a new class of black commercial farmers comes into tension with "its removal of subsidies and other support" which have resulted in "a uniquely hostile environment for new entrants into agriculture" (Hall 2015, 220; Movik 2012). This may be connected to the emphasis placed on economic development value of water as discourses of efficiency and productivity subsume values of 'equitable access'. The focus on promoting commercial farms contrasts with the deeply politicised initial aim of land reform under the RDP to provide secure tenure (Hall 2015, 137) and served values of rapid economic growth rather than transforming rural livelihoods and creating security for the poor during the post-apartheid policy process (Aliber and Cousins 2012, 145) and resulted in a lack of meaningful water reallocation (Tapela 2008).

Conclusion

South Africa is widely regarded as the first country in the world to have "promulgated national water legislation which uses water as a means to achieve societal transformation based on environmental and social justice" (Backeberg 2005, 17). However, this analysis has offered evidence that inequalities related to the realisation of economic value over equitable access to water have impeded the realisation of the reforming goals of the 1998 NWA. Based on the evidence presented from both literature review and fieldwork, this chapter has argued that the narrative of citizenship and rights to water, manifesting a value of 'equitable access', has been dominated, and undermined in practice, by a narrative of scarcity and the economic valuation of water use. This has resulted not just from competing sets of local interests, but also from the ways the post-apartheid state framed social and economic development in terms largely of private enterprise and withdrawal of state intervention. This not only generalised the logic and influence of commercial agencies but also struggled to build effective countervailing democratic institutions through which to promote a more effective political agenda of socially equitable approaches to water allocation.

The values-based approach taken in this chapter considered the power relations embedded in water governance and management as a contrast to alternative avenues that view water policy and practice from an apolitical techno-managerial approach. By examining the articulation of policy and practices in context, this analysis brings to the fore the underlying rationales, incentives, and power relations that influence the realisation of both stated and tacit values of water. A values-based approach to water, valuation, and power illuminates "constitutive processes and interconnections, thereby contributing to the production of concrete concepts" (Hart 2008, 15). By utilising a values-based approach to understand social relations and their relationship to forms of water governance and management, this research has sought to unravel the forces that socially produce inequities in water access and control (Movik 2012). This chapter has sought "to understand the complex relations between nature and society through a careful analysis of … the forms of access and control over resources and their implications for livelihoods" by explaining water struggles in terms of "knowledge, power, and practice" and "politics, justice, and governance" (Watts 2000, 257). It has argued that, contrary to the spirit of the NWA, white commercial farmers have been able to capture valuation techniques and employ the idea of scarcity to further exclude black people from water and land. Thus, despite the managerial, technological, and organisational principles guiding national government action related to water governance and management, it is ultimately the forms of political-economic power and its expression through values that shape mechanisms of access to and exclusion from water.

Notes

1 In early 2014 TSB was sold by its owner, Remgro group, to Rainbow Chicken Ltd (RCL), also owned by Remgro. It was renamed RCL Sugar.

2 A. Fischer, P. Mbiba and A. Emery, 'Onderberg Land Claims: for The State Attorney' (unpublished, 2010).

References

Aliber M. and Cousins B. (2012) "Livelihoods after land reform in South Africa" *Journal of Agrarian Change* 13: 140–165.

Backeberg G. (2005) "Water institutional reforms in South Africa" *Water Policy* 7: 107–123.

Bernstein H. (2013) "Commercial agriculture in South Africa since 1994: 'natural, simply capitalism'" *Journal of Agrarian Change* 13: 23–46.

Bernstein H. (2015) "Commercial farming and agribusiness in South Africa since 1994" in Cousins B. and Walker C. (eds) *Land divided land restored: land reform in South Africa for the 21st century* Jacana Media, Johannesburg, 104–119.

Bezuidenhout C. and Singels A. (2007) "Operational forecasting of South African sugarcane production: Part 1 – System description" *Agricultural Systems* 92: 23–38.

Brown J. (2011) "Assuming too much? Participatory water resource governance in South Africa" *Geographical Journal* 177: 171–185.

Brown J. (2013) "Can participation change the geography of water? Lessons from South Africa" *Annals of the Association of American Geographers* 103: 271–279.

Brown J. and Woodhouse P. (2006) "Pioneering redistributive regulatory reform. A study of implementation of a Catchment Management Agency for the Inkomati Water Management Area, South Africa" in Minogue M. and Cariño L. (eds) *Regulatory governance in developing countries* Edward Elgar, Cheltenham, 227–257.

Claassens A. (2015) "Law, land and custom, 1913–2014: what is at stake today?" in Cousins B. and Walker C. (eds) *Land divided land restored: land reform in South Africa for the 21st century* Jacana Media, Johannesburg, 68–84.

Cousins B. (2011) *What is a smallholder? Class-analytic perspectives on small-scale farming and agrarian reform in South Africa* Working paper 16, Institute for Poverty, Land and Agrarian Studies (PLAAS), University of the Western Cape.

Cousins B. and Walker C. (2015) *Land divided, land restored: land reform in South Africa for the 21st Century* Jacana Media, Johannesburg.

DWAF (2004) *National water resource strategy* Department of Water Affairs and Forestry, Pretoria.

DWAF (2005a) *Inkomati Water Management Area: internal strategic perspective* Department of Water Affairs and Forestry, Pretoria.

DWAF (2005b) *A draft position paper for water allocation reform in South Africa. Towards a framework for water allocation planning. Discussion Document* Department of Water Affairs and Forestry, Pretoria.

Faysse N. (2004) "Challenges for fruitful participation of smallholders in large-scale water resource management organisations: selected case studies in South Africa" *Agrekon* 43: 52–73.

Gelb S. (2007) "Macroeconomic policy in South Africa. From RDP through GEAR to ASGISA" in Gunnarsen G., Manus P.M., Nielson M. and Stolten H.E. (eds) *At the end of the rainbow? Social identity and welfare in the new South Africa* Southern Africa Contact, Copenhagen, 17–28 www.jakobsgaardstolten.dk/visitor/SACebook/SACebook(6).pdf

Hall R. (2015) "Who, what, where, how, why? The many disagreements about land redistribution in South Africa" in Cousins B. and Walker C. (eds) *Land divided, land restored: land reform in South Africa for the 21st Century* Jacana Media, Johannesburg, 130–148.

Hart G. (2008) "The provocations of neo-liberalism: contesting the nation and liberation after apartheid" *Antipode* 40: 678–705.

Hart G. (2010) "D/developments after the meltdown" *Antipode* 41: 117–141.

Hassan R. and Farolfi S. (2005) "Water value, resource rent recovery and economic welfare cost of environmental protection: a water-sector model for the Steelpoort sub-basin in South Africa" *Water SA* 31(1): 9–16.

Hope R. and Garrod G. (2004) "Household preferences to water policy interventions in rural South Africa" *Department for International Development* 6: 487–499https://www.gov.uk/dfid-research-outputs/household-preference-to-water-policy-interventions-in-rural-south-africa (Accessed 3 February 2018).

James P. and Woodhouse P. (2017) "Crisis and differentiation among small-scale sugarcane growers in Nkomazi, South Africa" *Journal of Southern African Studies* 43: 535–549 www.tandfonline.com/doi/full/10.1080/03057070.2016.1197694 (Accessed 3 February 2018).

Kemerink J., Ahlers R. and van der Zaag P. (2011) "Contested water rights in post-apartheid South Africa: the struggle for water at catchment level" *Water SA* 37: 585–594.

King B. (2007) "Developing KaNgwane: geographies of segregation and integration in the new South Africa" *Geographical Journal* 173: 13–25.

King B. and McCusker B. (2007) "Environment and development in the former South African Bantustans" *Geographical Journal* 173: 6–12.

Lankford B., Bakker B.K., Zeitoun M. and Conway D. (eds) (2014) *Water security: principles, perspectives, practice* Routledge, London.

Malzbender D., Goldin J., Turton A. and Earle A. (2005) *Traditional water governance and South Africa's National Water Act: tensions or cooperation?* International workshop on 'African water laws: plural legislative frameworks for rural water management in Africa' 26–28 January, Johannesburg, South Africa.

Meissner R., Funke N. and Neinaber S. (2013) "The status quo of research on South Africa's water resource management institutions" *Water South Africa* 39: 721–732.

Merrey D. (2011) "Time to reform South African water reforms: book review in water alternatives" in Schreiner B. and Hassan R. (eds) *Transforming water management in South Africa: designing and implementing a new policy framework*. Global Issues in Water Policy, Volume 2. Springer, Dordrecht.

Movik S. (2012) *Fluid rights: South Africa's water allocation reform* HSRC Press, Cape Town.

Movik S. (2014) "A fair share? Perceptions of justice in South Africa's water allocation reform policy" *Geoforum* 54: 187–195.

Muller M. (2007) "Parish pump politics: the politics of water supply in South Africa" *Progress in Development Studies* 7: 33–45.

Muller M. (2008) "The challenges of implementing an African water resource management agenda" in Kondlo K. and Ejiogu C. (eds) *Africa in focus: governance in the 21st century* HSRC Press, Pretoria, 146–162.

Nieuwoudt W., Backeberg G. and Du Plessis H. (2004) "The value of water in the South African economy: some implications" *Agrekon* 43: 162–183.

Peet R. (2002) "Ideology, discourse, and the geography of hegemony: from socialist to neoliberal development in post-apartheid South Africa" *Antipode* 34: 54–84.

Peters R. (2015) "*Some are more equal than others": the prioritization of development over equity values in South Africa's water reform.* MSc dissertation, University of Manchester.

Pollard P. and Du Toit, D. (2011) "Towards adaptive integrated water resources management in southern Africa: the role of self-organisation and multi-scale feedbacks for learning and responsiveness in the Letaba and Crocodile catchments" *Water Resources Management* 25: 4019–4035.

Postero N. (2007) *Now we are citizens: indigenous politics in postmulticultural Bolivia* Stanford University Press, Stanford, CA.

Schreiner B. (2013) "Viewpoint – why has the South African National Water Act been so difficult to implement ? *Water Alternatives* 6: 239–245.

Schreiner B. and Van Koppen B. (2002) "Catchment management agencies for poverty eradication in South Africa" *Physics and Chemistry of the Earth* 27: 969–976.

Schreiner B. and Hassan R. (eds) (2011) *Transforming water management in South Africa: designing and implementing a new policy framework*, Global Issues in Water Policy, Volume 2. Springer, Dordrecht.

Seshoka J., Lange W. and Faysse N. (2004) "The transformation of irrigation boards into water user associations in South Africa: case studies of the Lower Olifants, Great Letaba and Vaalharts WUA Water User Associations" *International Water Management Institute* 72: 1–64.

Swyngedouw E. (2006) *Human development report: power, water, and money – exploring the nexus* United Nations Development Programme, Oxford University Center for the Environment, Oxford http://hdr.undp.org/en/content/power-water-and-money-exploring-nexus (Accessed 3 February 2018).

Tapela B. (2008) "Livelihoods in the wake of agricultural commercialization in South Africa's poverty nodes: insights from small-scale irrigation schemes in Limpopo Province" *Development Southern Africa* 25: 181–198.

Tewari D. and Oumar S. (2013) "Is the water permit system a panacea or a bed of inefficiency? the case of South Africa" *Water Policy* 15: 570–584.

Turton A.R. (2000) "Precipitation, people, pipelines and power in southern Africa: towards a 'virtual water'-based political ecology discourse" in Stott P. and Sullivan S. (eds) *Political ecology: science, myth and power* Edward Arnold, London, 132–153.

Van Koppen B. (2008) "Redressing inequities of the past from a historical perspective: the case of the Olifants basin, South Africa" *Water SA* 34: 432–438.

Van Koppen B., Jha N. and Merrey D. (2002) "Redressing racial inequities through water law in South Africa: interaction and contest among legal frameworks" in Pradhan R. (ed.) *Legal pluralism and unofficial law in social, economic and political development* Proc. XIIIth International Congress of the Commission on Folk Law and Legal Pluralism 7–10 April, Chiang Mai, Thailand. Volume 2: International Centre for the Study of Nature, Environment and Culture, Kathmandu, Nepal, 201–219.

Von Schnitzler A. (2008) "Citizenship prepaid: water, calculability, and technopolitics in South Africa" *Journal of Southern African Studies* 34: 899–917.

Watts M. (2000) "Political ecology" in Sheppard, Eric, and Barnes, Trevor J. (eds) *A companion to economic geography* Wiley Blackwell, Oxford, 257–274 http://onlinelibrary.wiley.com/book/10.1002/9781405166430

Weeks J. (1999) "Stuck in low GEAR? Macroeconomic policy in South Africa 1996–1998" *Cambridge Journal of Economics* 23: 795–811.

Woodhouse P. (1995) "Water rights and restructuring in South Africa: a case study from Eastern Transvaal" *Water Resources Development* 11: 527–544.

Woodhouse P. (2012) "Reforming land and water rights in South Africa" *Development and Change* 43: 847–868.

Woodhouse P. and Hassan R. (1999) Implementation of the National Water Act. Catchment Management Agencies: Interests, Access and Efficiency. Inkomati Basin Pilot Study prepared for DWAF, South Africa and DIFD-Southern Africa, UK. http://hummedia.manchester.ac.uk/institutes/gdi/publications/workingpapers/archive/rr/rr_wp12.pdf

Young I. (1990) *Justice and the politics of difference* Princeton University Press, Princeton, NJ.

Zeitoun M., Warner J., Mirumachi N., Matthews N., McLaughlin K., Woodhouse M., Cascão, A. and Allan J.A. (2014) "Transboundary water justice: a combined reading for analysis and diplomacy of literature on critical transboundary water interaction and 'Justice'" *Water Policy* 16: 174–193.

12 Conclusion

The limits of economic valuation

Sarah Bracking, Aurora Fredriksen, Sian Sullivan and Philip Woodhouse

The chapters collected in this volume each explore instances where economic valuation systems are being newly proposed, assembled and/or deployed in the fields of development, environment and conservation. While the valuation practices of different cases range widely in form – from the formalised financial instruments of carbon markets, to the fluid qualitative assessments performing 'value for money' at UK DfID, to the many more or less explicit economic valuation systems in between – one feature is strikingly common: these newly economic systems are often not working in the manner in which they were designed. At least, they are not working in the smooth, self-contained way imagined by both proponents and opponents of establishing the relative worth of things through economic valuations. It is not that these economic valuations necessarily fail outright (although they come close to this in Chapter 7 on DEFRA's biodiversity offsetting pilot), but rather that they fail to work as intended, variously requiring continuous work to function, incorporating 'patches' and 'fixes', and/or encountering resistance in unexpected places.

In this concluding chapter, we consider the limits to economic valuation in development, environment and conservation, pulling out a series of intersecting points where these limits become legible in our case studies: the limits of commensurability, of measurement and measurability, and of stakeholder buy-in and subsequent attempts to contain resistance to these new economic orderings. We also explore how and why these limits to economic valuation have not led to abandoning aims to value things in economic terms, but rather to renewed, iterative attempts to impose orders of worth defined through economic values. This is sometimes due to limitations in the imaginations of a planning or expert community to do something in a different way, but it is also closely linked to the strong correlation between the interests of money-holders, and the pursuit of economic calculations in human affairs.

In other words, it is often power struggles that characterise the context of these iterative experiments and which condition their development, and these, in turn, reflect the structures of power in society more broadly. From these structures also emerge concentrations of agency that often provide resistance to valuation processes. These include the humans, non-humans and things that overflow or exceed a value-making assemblage, such as those who wish to do value calculations in alternative ways. These are the communities who perhaps have traditionally been seen as outside capitalist processes. But resistances and participations come from both outside *and* inside the institutional assemblage of a valuation process, and expert knowledge-holders also assert agency in challenging and changing calculative devices and processes (as was demonstrated in our chapters on carbon, biodiversity, infrastructure and water). There is also a coalition of agents who advocate for economic valuations to dominate evaluative situations, such as investors and capital-holders, who often experience a reward structure when these calculations are implemented in this way (as described in Chapter 6 for impact investing in conservation finance), instead of, and in place of, other forms of valuation.

Setting the case studies in the preceding chapters in the context of a global political economy, then, we interrogate the power relations at work within and across them. It is these power relations, we argue, that work to support the continued advancement of economic valuations in new domains, despite the limits of calculability and commensuration. At the same time, we argue that attending to the limits of economic valuations is worthwhile, as these limits may be leveraged in struggles against the progressive economisation of worth in the fields of development, environment and conservation.

Limits to commensurability

The first and most obvious source of limits to economic valuation is the difficulty in achieving commensurability between valued entities. Efforts to calculate the worth of something are constantly and iteratively experimented, never really working, but never really failing either, and frequently requiring new calculations whose representations of worth are continuously negotiated and often contested. This search for calculative devices is illustrated in the biodiversity offsetting case history worked through in Chapter 7 and also in the efforts to make air pollution with greenhouse gases tradable in units of carbon equivalent. Elaborating a scientific base from which to make different gases commensurable was an early experiment that underpinned the UNFCCC 'black box' of conversion factors used to transform diverse types of emissions into tonnes of carbon equivalent (MacKenzie 2008; Ehrenstein and Muniesa 2013). This was initially a successful fix, creating a firewall around the calculations of commensurability, despite many critics of the science on which they were based. However, creating an actual carbon offset commodity in place proved much more unstable, as Watt shows in Chapter 8, since the tonnes of carbon equivalent had to be priced in relation to physical factories or forests, and the changes ('additionality') achieved in measurable emissions. This case showed that even if a relative fixity of carbon calculation was initially achieved, once commensurability has to extend across sites, factories, mines, forests and so forth to enable trade, the system lacked the degree of measurability and trust required to uphold a market.

The case of the Green Climate Fund, as discussed by Bruun in Chapter 9, illustrates how this problem of commensurability across different climate change projects to be selected for funding was contested by the participants involved in the institutional assemblage. Bruun argued that in the design of the Green Climate Fund, two meta-discourses were prevalent within the UNFCCC process: green governmentality and ecological modernisation. These worked in conjunction to enforce an economic calculus in the GCF design, and thus undermined values drawing on a third meta-discourse on civic environmentalism presented by the Civil Society Organisations on human rights and pro-poor climate finance. Here the participants in the assemblage argued over how the value calculations within the Green Climate Fund would be carried out. At the holistic level, these were contests about what should be funded according to pre-existing preferences of participants for energy, infrastructure, REDD + and so forth. But at a more micro-level they also became contests over how benefit would be represented, which in turn would inform future investment decisions. In this contest over how the valuation process would be designed, some knowledge and services provided by CSOs emerged as relevant, while others were deemed discountable. Arguably, the calculative processes contained in defining 'Green', and in describing the benefits from investments, brought the CSO community closer to the banking community in terms of how they selected, processed and gave worth to different types of knowledge. The Green Climate Fund is now functional and uses two metrics to show its worth: carbon emissions avoided; and beneficiaries – defined as the "anticipated number of people with increased resilience". While appearing initially to be quite simple metrics, both contain profound problems of commensurability in practice: the first because of the problems in measuring avoidance and additionality, discussed also by Watts; and the second, because of the nebulous meaning of resilience.

So far, therefore, each round of engagement between the different groups involved in negotiating the Green Climate Fund tends to end in an approximation which is contested, and then met with another effort in measurement. This latter effort usually acts to restore the authority of the experts and professionals within the organisation in the face of resistance from the NGOs, who struggle again to balance the financial imperatives of a GCF with their human-centred concerns with ecological justice and human welfare. What is being performed here, in arguments about how to carry out the valuation of climate change mitigation efforts, is a power struggle between different stakeholders in the process of climate change. The power struggles beyond the STA are thus also represented within it, and often the types of calculation chosen tend to favour the expertise and authority of the capital-holders within the GCF. While the diversity of the portfolio of projects actually funded would defy most logics of commensurability, the act of imposing it, however inaccurately, brings all these people, non-humans and things, within the governance orbit of the bankers and governments who liquidate the GCF. This, of course, is never the end of the experimental story, since further overflows and inconsistencies will inevitably emerge.

The limits to commensurability in valuation processes can be seen in all our case studies, either in terms of pacifying the thing which is to be commodified, such as habitats or species, or in terms of viewing the same thing as commensurate in different places and contexts, such as water. In Woodhouse and Muller, Chapter 10, for example, the authors show how efforts to impose uniform values on water are confounded by it acting differently in different places, with different users valuing different aspects of its agency – depth, flow, volume, quality – depending on the context. This confounds commensurability across places or scales – even of units of the same commodity, let alone that economic valuation often seeks commensurability for units of different things! Similarly, we saw in Peters and Woodhouse, Chapter 11, how white commercial farmers were able to subvert legislation drafted to promote redistribution of water rights and instead use a category of existing legal use to maintain their privilege. Over and above the problems of establishing commensurability of the valued units, this example shows that power relationships, which are in some senses outside the STA as well as dominant within it, can also confound efforts at uniform or standardised valuation. In other words, the commercial farmers were able to insist that their access to water was more important than that of other peoples, whatever commensurability might be established for the unit of water itself.

Limits to measurement and measurability

Limits to commensurability are in part an outcome of the limits to measurement and measurability arising from the unruliness of externalities being drawn into domains of formal economic valuation in our case studies. In other words, they are due to the inherent vitality of people and things. Conjuring a 'biodiversity unit' into being, for example, discounts the dynamic relational ecologies that permit its existence, arguably creating an anti-ecological hybrid value entity that nonetheless is intended to display and hold value (as considered in Sullivan's and Carver's work in Chapters 6 and 7). Meanwhile, the case studies within the development domain illustrate the challenges posed to measurement particularly well, since human health and happiness are inevitably messy and chaotic. For example, in Chapter 3, Fredriksen shows how the political imperative to impose a value for money metric within the UK DfID – a highly standardised and pacifying economic device – brought conflicts with pre-existing and emerging alternative world views of valuation within development. In Chapter 4, Machingura describes how those infected or affected by HIV prove hard to capture by the metrics of life-years favoured by health service managers.

Instead of creating certainty, precision and predictability, what we see instead in our cases, then, is a proliferation of techniques of measurement and valuation, as different users of calculative devices vie for ownership of the generalisable and dominant calculative device. For example, in the case of DfID's calculation of VfM, various aid programmes and projects were reliant on proving VfM in order to secure further funding, such that many agents emerged to compete over the meaning and calculation of VfM. This is not unusual in organisations when a new calculative device emerges and the calculative method to populate it comes later and has consequences for funding distributions. Empirical research thus demonstrates the ongoing, iterative and diversifying struggles over the design and application of calculative devices intended to fabricate and stabilise value (as shown in Chapter 7). These struggles may involve amendments that are poorly suited as representational calculations but still come to substitute for the relational entities being 'valued' through arithmetic applications (see discussion in Sullivan and Hannis 2017). Value may nonetheless accrue to the figures arising from these proliferating numerical applications, albeit poorly coupled with the material worlds being thus represented.

Limits to containing resistance

As well as limits to commensurability and buy-in, the complexities of the more-than-human objects of valuation continue to defy economic calculability in many of the cases explored here. In these cases the limits to economic valuation are not the competency of the actor, or the complexity of the subject, but that calculability itself has real limits based on what is available to maths and science. For example, the efforts to forestall environmental catastrophe by anticipating the most dangerous point of global warming and then working back to avoid it – perhaps not the safest methodology in the first instance – involve warming models and carbon calculations that are at the edge of modern science. Or the attempt to conserve species by merely protecting the lowest sustainable number for breeding purposes – may present itself as the least cost solution, but remains risky if the natural science calculations of probable breeding potentials fall short. Similarly, in the case of the plan for Durban's port expansion (Chapter 5), the rationality of the economic proposals relied on predictions of growth in container shipping, and predictions of employment creation to justify evicting a community. However, at a closer glance, even these economic calculations were fallible and consequently subject to successful resistance by the affected communities.

The challenge for public policy is thus to protect and conserve without necessarily relying on quantitative calculability, which could involve error despite its apparent rationality. One alternative means of doing this is by imposing orders of moral worth, which is particularly challenging in our contemporary financialised culture, when there is no obvious profit-making opportunity in the course of the social action to be taken. Yet it is also clear that public policy goals can appeal to social norms of worth that are not quantified, let alone monetised, if there is sufficient community agency to insist on this form of valuation. This reflects the sense of people making decisions as citizens rather than as consumers. It also highlights the sense that people are not human 'items' but defined by their social relationships with other people. In these, as Machingura's chapter (4) makes clear, qualities such as 'respect' are of the greatest value.

Power in the global political economy

Despite the variety of limits to economic valuation illustrated in the case studies collected here, efforts to assign worth through economic valuation have not been abandoned in the fields of development, environment and conservation. Indeed, the limits of economic valuation often seem to only encourage further iterations of economic values and economising practices. There are many cases, for example, where the act of measurement is its most crucial element, whatever the data or figure or category that emerges. That it is done at all is the most important feature. The act of measurement has become an end in itself, because it functions as legitimation not only of the measurement but of the discourse that has prescribed it as necessary. In other words, the act of measurement itself becomes the end goal, almost regardless of the consequences of each individual act of measurement, in a form of zombie neoliberalism (Peck 2010). In this, the mindless imposition of an austerity-driven neoliberalism leaves those state functionaries charged with welfarism – the 'left hand of the state' – the 'sense that the right hand no longer knows, or worse, no longer really wants to know what the left hand does' (Bourdieu 1998 [1992], 2). In this duality from Bourdieu, the right-hand of the state is made up of the coercive and fiscal authorities intent on imposing a vision of austerity into a political project of control with all its contradictory tendencies (see Peck 2010, 104). While Wacquant (2009) insists, drawing on this work, that the 'left' and 'right' hands of the state are working in a symbiotic way that is complementary in the making of the neoliberal subject, the evidence collected here confirms rather the original imaginary of Bourdieu, of more contested spaces and contradictory processes.

We have illustrated in this book, in the context of the global political economy, that the powerful can work with measurement to their advantage, either by subverting its meaning, using its categories, headlining its failures or supplanting alternative methodologies. We also have evidence of this process being disrupted, disturbed and stalled by other agents, including the vulnerable and those adversely affected. For the most part, however, those who instigate, design and implement valuation processes tend to be the more powerful agents in an assemblage. For those individuals or organisations that can successfully launch an index, measurement system, equation, software platform, proprietary information stream or some such, there are opportunities for authority, expertism and ultimately profit. The political economy of a global capitalism rewards calculation, and the more opaque and complex-looking the better.

In terms of the worth to health in Zimbabwe, for example, internal measurement of living years was important to the internal performance of the Ministry of Health, but had little carry outside the organisation. Similarly, measures of the economic value of water were used in South Africa to prioritise irrigation by commercial farmers over increasing water availability for the largely black, poor population. Economic value determines that providing water to commercial farmers constitutes a benefit, while providing tapped water to the poor was framed as a cost. Similarly, providing further berths to international shipping was framed as a benefit, while maintaining the survival and existence of people and their small businesses in the 'back of port interface zone' was framed as a cost.

Thus the performativity of measurement is a component of power and authority. In this, measurement exists in a semi-autonomous sense to both that which is measured and the reliability or accuracy of the measurement effort. That so much effort is invested in successive and iterative experiments in calculability is related to how power and authority are justified in capitalist systems more generally, and, while this is a complex relationship, governance is largely exercised successfully, and with economic reward, by those who have measurability at their command.

From economic valuation to political contestation

Overall, in placing the making of new economic values in the calculative, discursive and institutional contexts of their manifestation, our volume affirms the "socially contingent and radically open character of value(s)" (Weeks 1993, 195). Simultaneously, the cases read together illuminate the neoliberal, and thus ideological, calculations of worth crossing development, environmental and conservation domains of value. In inventing and consolidating a particular truth of value, this position closes possibilities both for future alternative universal solidarities of worth, as well as for value frames and practices that become other(ed) in the process.

However, our cases also illustrate the fragility of economic valuation in establishing value and worth, particularly in relation to their economic and neoliberal manifestations. None of our case studies serves to establish the success of economics, or financial valuation, in a permanent sense. Indeed, some, such as the CDM and biodiversity offsetting, arguably mostly fail. This might reflect the weakening of neoliberalism globally, although it is beyond the scope of this volume to determine whether neoliberalism as a project is strengthening or weakening. At the level of our case studies, however, the various processes of valuation share a governance feature, in that they all change the mode or technology of power in use: they depoliticise a particular issue by rendering its logic and calculability into an exclusively technical question. In other words, valuation processes tend to frame an issue in such a way that explicit political contestation, which could be expressed in deliberative argument and discussion, is removed from the context. On the other hand, if a group of affected persons can escape an STA that renders political problems as 'technical' this moves decision-making into a more explicitly political arena. The quality of decisions about what we value then becomes – for better or worse – a function of the nature of political discourse. Moreover, if 'rendering technical' serves to mask the exercise of power, then a more explicitly political discussion of values will not prevent the exercise of power, but make it more apparent who is doing what to whom. While this will improve the quality of public debate, a more transparent exercise of power may, in turn, improve the technical and calculative quality of processes and devices within valuation assemblages.

References

Bourdieu, P. (1998) "The left hand and the right hand of the state", in P. Bourdieu *Acts of Resistance*, pp. 1–10. Cambridge: Polity.

Ehrenstein, V. and Muniesa, F. (2013) "The conditional sink: counterfactual display in the valuation of a carbon off setting restoration project", *Valuation Studies* 1(2): 161–188.

Laclau, E. (1989) *New Reflections on the Revolution of Our Time*. London: Verso.

MacKenzie, D. (2008) "Making things the same: gases, emission rights and the politics of carbon markets", *Accounting, Organizations and Society* 34(3–4): 440–455.

Peck, (2010) "Zombie neoliberalism and the ambidextrous state", *Theoretical Criminology* 14(1): 104–110.

Sullivan, S. and Hannis, M. (2017) "'Mathematics maybe, but not money': on balance sheets, numbers and nature in ecological accounting", *Accounting, Auditing and Accountability Journal* 30(7): 1459–1480.

Wacquant, L. (2009) *Punishing the Poor: The Neoliberal Government of Insecurity*. Durham, NC: Duke University Press.

Weeks, J. (1993) "Rediscovering values", in Squires, J. (ed.) *Principled Positions*, pp. 189–211. London: Lawrence and Wishart.

Index

Note: Page numbers in **bold** refer to tables and page numbers in *italics* refer to figures.

For Product Safety Concerns and Information please contact our EU representative GPSR@taylorandfrancis.com Taylor & Francis Verlag GmbH, Kaufingerstraße 24, 80331 München, Germany